T Lymphocytes
Structure, Functions, Choices

NATO ASI Series

Advanced Science Institutes Series

A series presenting the results of activities sponsored by the NATO Science Committee, which aims at the dissemination of advanced scientific and technological knowledge, with a view to strengthening links between scientific communities.

The series is published by an international board of publishers in conjunction with the NATO Scientific Affairs Division

A	Life Sciences	Plenum Publishing Corporation
B	Physics	New York and London
C	Mathematical and Physical Sciences	Kluwer Academic Publishers
D	Behavioral and Social Sciences	Dordrecht, Boston, and London
E	Applied Sciences	
F	Computer and Systems Sciences	Springer-Verlag
G	Ecological Sciences	Berlin, Heidelberg, New York, London,
H	Cell Biology	Paris, Tokyo, Hong Kong, and Barcelona
I	Global Environmental Change	

Recent Volumes in this Series

Series A: Life Sciences

T Lymphocytes

Structure, Functions, Choices

Edited by

Franco Celada

The Hospital for Joint Diseases, New York University
New York, New York

and

Benvenuto Pernis

Columbia University
New York, New York

Springer Science+Business Media, LLC

Proceedings of a NATO Advanced Study Institute
on T Lymphocytes: Structure, Functions, Choices,
held September 15–27, 1991,
in Porto Conte (Alghero), Sardinia, Italy

NATO-PCO-DATA BASE

The electronic index to the NATO ASI Series provides full bibliographical references (with key-words and/or abstracts) to more than 30,000 contributions from international scientists published in all sections of the NATO ASI Series. Access to the NATO-PCO-DATA BASE is possible in two ways:

—via online FILE 128 (NATO-PCO-DATA BASE) hosted by ESRIN, Via Galileo Galilei, I-00044 Frascati, Italy.

Additional material to this book can be downloaded from http://extra.springer.com.

Library of Congress Cataloging-in-Publication Data

T Lymphocytes : structure, functions, choices / edited by Franco
Celada and Benvenuto Pernis.
      p.   cm. -- (NATO ASI series. Series A, Life sciences ; vol.
233)
      "Proceedings of a NATO Advanceed Study Institute on T Lymphocytes:
Structure, Functions, Choices, held September 15-27, 1991, in Porto
Conte (Alghero), Sardinia, Italy."--T.p. verso.
      "Published in cooperation with NATO Scientific Affairs Division."
      Includes bibliographical references and index.
      ISBN 978-1-4613-6332-3   ISBN 978-1-4615-3054-1 (eBook)
      DOI 10.1007/978-1-4615-3054-1
      1. T cells--Congresses.   I. Celada, Franco. II. Pernis,
Benvenuto. III. North Atlantic Treaty Organization. Scientific
Affairs Division.   IV. NATO Advanced Study Institute on T
Lymphocytes: Structure, Functions, Choices (1991 : Porto Conte,
Italy) V. Series: NATO ASI series. Series A, Life sciences ; v.
233.
      [DNLM: 1. T-Lymphocytes--physiology--congresses.   WH 200 T1114
1991]
QR185.8.T2T2   1992
599'.029--dc20
DNLM/DLC
for Library of Congress   92-49284
                                                               CIP

ISBN 978-1-4613-6332-3

COMMITTEE

Directors:	Franco Celada and Benvenuto Pernis
Secretary:	Pierluigi Fiori
Scientific Committee:	K. Rajewsky, E. Sercarz, M. Siniscalco
Faculty Members:	O. Acuto, L. Adorini, J. Bodmer, W. Bodmer, C. Bona, H. Cantor, F. Celada, I. Cohen, K. Hannestad, E. Heber-Katz, K. Karjalainen, A. Lanzavecchia, F. Manca, D. Mathis, N. A. Mitchison, S. Nathenson, D. Parker, B. Pernis, E. Sercarz, A. Theofilopoulos, T. Tada, H. von Boehmer, M. Zanetti
Local Staff:	Paola Melis, Paola Rappelli

PREFACE

This volume contains the proceedings of the NATO Advanced Study Institute held in Porto Conte (Alghero), Sardinia, September 15-27, 1991.

The A.S.I. was attended by 86 graduate and postgraduate students from 18 different countries, and was hosted by the newly founded International Laboratory of Molecular Genetics of Porto Conte, directed by Prof. Marcello Siniscalco. The A.S.I. was funded by NATO Scientific Affairs Division, the International Union of Immunological Societies, the European Community (Directorate General for Science, Research and Development), the Italian Research Council, and the San Raffaele Institute of Milano. In addition, a number of students who reside in the U.S. received travel funds from the U.S. National Science Foundation, and the Turkish National Fund provided financial assistance to several students from Turkey.

When we decided to organize a course on T lymphocytes, our concern was to reach a balance between the teaching of both the hardcore principles and the latest experimental findings of cellular immunology, and the recently expanded interfaces with the not-yet-known: hypotheses, speculations, new projections to be born from the discussions. *A posteriori*, we believe we have largely succeeded in this endeavor, or, to use a more objective statement, that the course turned out to be a balanced mixture of teaching and contributing, of learning and discussing as we had hoped, and that this happened by a number of mechanisms and factors for only a fraction of which we can claim responsibility or merit. Yes, we were responsible for the selection of the members of the scientific committee, and together with them for the choice of the faculty. Yes, we had the merit to select a venue which was ideal in terms of sheer beauty, but also relative isolation, and a relation (at least mental) to a new research institute in the labor of being born. Yes, the capable secretary of the course, who also brought an extraordinary local staff, was our choice. But what about the weather (twelve days without faltering, whose influence on everybody's mood was visible), the quality of the students and their willingness to ask, talk, engage? And what about the theorists of immunology, a small group of mathematicians and a physicist who came as observers, did not miss one session, and contributed constantly to the discussion? We hope that some of the flavor of these twelve days in Porto Conte will be preserved in this book. It contains 16 chapters written by faculty members and 8 by students who had contributed a presentation at the course.

The basic theme of the course was the structure and the function of T lymphocytes. These cells are rightly considered to be the commanding elements of the immune system. In fact the immune response to protein antigens, those for which the distinction between self and non-self is most difficult and most important, is decided by T cells. For these antigens, B cells require the "help" or, more precisely, the "permission" of T cells, and not the other way around. With this premise, it is not surprising that the structure and the function of T cells are particularly complex. To begin with, the T cell receptors do not recognize intact protein antigens but rather fragments of these, that have been broken down to the level of peptides in the antigen-presenting cells and are presented to the T cells after having been bound into a specific cleft of membrane molecules, the histocompatibility antigens. This process is grounded on clonal selection and is the basis of immunological tolerance. T cells are also involved in intercellular regulations that give to the immune system its capacity to respond in an integrated way with the maximum efficiency and an (inevitable) minimum of potentiality to inflict damage to the organism itself.

Thus, the complexity of the immune system is to a large extent related to the complexity of the T-lymphocyte system, which is not likely to yield soon to a comprehensive analytical approach, including the exciting efforts of computerized simulations. The study of that system is, without doubt, at the cutting edge of our capacity to understand, and perhaps control, some major plagues that still affect the human species, like AIDS, and the major autoimmune diseases like multiple sclerosis, rheumatoid arthritis and juvenile diabetes. These underlying practical problems, as well as the intellectual challenge posed by the T system, explain the interest and excitement that was obvious in the faculty and students throughout the course.

F. Celada
B. Pernis

CONTENTS

T-CELL ANTIGEN AND MHC RECOGNITION:
MOLECULAR ANALYSIS OF HUMAN α/β TCR SPECIFIC
FOR A TETANUS TOXIN-DERIVED PEPTIDE

Brigitte Boitel, Myriam Ermonval, Ulrich Blank
and Oreste Acuto

Laboratory of Molecular Immunology
Department of Immunology, Pasteur Institute
25, rue du Dr. Roux, 75724 Paris cedex 15, France

INTRODUCTION

Most immune responses are initiated through recognition of foreign antigen, associated with self MHC-encoded products, by T-cells via a clonotypic cell surface receptor (T-cell antigen receptor, TCR) (1). Much evidence has accumulated during the past few years indicating that the TCR recognizes short peptide fragments of intracellularly degraded antigens laying in a groove in the most external domain of class I or class II MHC-encoded molecules (2-4). Therefore, in contrast to B-cells whose Ig antigen receptors can directly interact with native proteins, the TCR forms a trimolecular complex with peptide and MHC. The TCR is a disulfide-linked heterodimer composed of two transmembrane chains, α (~50 kDa) and β (~40 kDa), each containing an amino terminal variable domain and a carboxy terminal constant domain (5). The α/β dimer is non-covalently associated at the cell surface with at least six smaller transmembrane subunits, γ, δ, ε2 and ζ2 (or ζη) responsible for coupling antigen recognition to the T-cell activation pathways (6, 7). The TCR specificity for antigen/MHC resides in the amino terminal domains of the α and β chains which are generated, like Ig, as a result of combinatorial juxtaposition of germline-encoded variable (V), diversity (D) (for the β chain) and junctional (J) segments mediated by site-specific recombination mechanisms during T-cell differentiation. For a review on the genomic organization of α and β loci and rearrangement mechanisms see references (5, 8).

T Lymphocytes: Structure, Function, Choices, Edited by F. Celada
and B. Pernis, Plenum Press, New York, 1992

About 100 Vα and 60 Jα segments, many of which have been isolated and sequenced in both human and mouse (8, 9, 10), are thought to exist. Approximately thirty mouse Vβ (11) and 57 human Vβ (12) gene segments have been described and in both species two Dβ and 12-13 functional Jβ segments are found (5, 13). Based on amino acid sequence similarity, Vα and Vβ germline segments can be grouped into 20-30 subfamilies generally composed of one to several members (8-12, 14). In addition to combinatorial assortment of germline V, D and J segments, imprecise Vβ to Dβ, Dβ to Jβ and Vα to Jα joining and N-region additions in both α and β chains dramatically contribute to increase the diversity of the TCR variable domains (5). Thus, α and β junctional regions can be found to vary considerably in length, (~20 to 27 residues can be found to connect the V segment to the constant region), and amino acid composition (5, 15). However, in contrast to Ig, somatic mutations in the rearranged TCR V gene segments do not appear to play a significant role in the generation of diversity (16), implying perhaps that preservation of the germline-encoded sequences is an important factor in maintaining TCR specificity. In spite of this apparent limitation, the contribution of both combinatorial association of germline-encoded gene segments and junctional diversity can be expected to generate a repertoire of approximately 10^{14}-10^{15} different α/β TCR in a single individual (5).

Amino acid sequence comparison of the α and β V regions with their Ig counterpart has shown a remarkable conservation of residues critical for maintaining the basic architecture of the Ig V regions (17, 18). Thus, although the three-dimensional structure of the TCR is not yet known, it is predicted that its V regions fold and pair similarly to Ig V regions. Furthermore the identification of hypervariable regions in both α and β subunits (19-21) (more evident, however, for the β chain) at sites corresponding approximately to the Ig complementarity determining regions (CDR) 1, 2 and 3, the latter being the result of junctional variability, suggests that in the TCR these regions contribute to form the binding site interacting with the peptide/MHC complex. Accordingly, differences in the peptide/MHC specificity of TCR appear to correlate with amino acid changes in the putative CDR (15, 22-27). Moreover, some studies have shown that point mutations introduced in these regions may substantially alter TCR recognition (28, 29).

Why do Ig and TCR, whose molecular architecture and strategy to generate variability in their respective binding sites appear to be so similar, recognize antigens in such a drastically different way? Does the TCR possess structural features, not obvious from the primary structure nor from the modeling on Ig V regions, which determine its capacity to interact with MHC molecules? And if so, are there sites in the TCR variable regions specialized to interact with the peptide?

Clearly, knowledge of the three-dimensional structure of the TCR is an essential step to answer these questions. In this respect, recent progress in generating soluble α/β TCR (30, 31 and K. Karjalainen, personal communication) is very encouraging. Moreover, it has been demonstrated that soluble class I and class II molecules carrying a single peptide can be isolated (32, 33). In addition to physico-chemical studies on the binding of TCR to peptide/MHC, this should allow co-crystallisation of tri-molecular complexes to carry out detailed structural analysis of their interaction.

A complementary approach, which may give key information on the structural basis of TCR recognition, is to correlate the primary structure of TCR specific for defined peptide/MHC complexes with their fine specificity by using substituted analogs of the peptide (22-27) and site-directed mutants of MHC molecules or, whenever possible, different MHC alleles presenting the same peptide (15, 22). In addition, to provide clues about the residues of the peptide and MHC contacting the TCR, the combination of these approaches may indicate which structural components of the TCR V regions contact the antigen and/or the MHC. It should be then possible to test more directly the existance of such contacts by manipulating the structure of the putative CDR by introducing site-directed mutations (28, 29).

Primary structure analysis of TCR specific for defined peptide/MHC class II and class I complexes in the mouse have indicated that receptors with the same specificity tend to use a limited set of V gene segments (22-27, 33-40) and that the use of particular Vα and/or Vβ regions may correlate with antigen or MHC recognition (22-25, 27, 28, 34-39, 41). Moreover, some studies have shown that TCR using the same germline V gene segments often express very similar (or identical) junctional sequences both in length and amino acid composition, suggesting that V regions and junctional sequences are functionally coselected (24-26, 33, 36-37).

An elegant approach, which made use of single α or β transgenic mice, has provided compelling evidence that in the Ia-restricted response to an immunodominant cytochrome c-derived peptide, amino acid residues invariably found at defined positions in the junctional regions of both α and β chains (expressing identical V gene segments) directly contact distinct side chains of the peptide (33). These data support models of the TCR/peptide/MHC interaction in which the junctional regions would lay above the MHC groove and would contact the peptide while the MHC would be essentially interacting with the putative (less variable) CDR1 and CDR2 encoded by germline V gene segments (5, 42). The latter would have the ability of recognizing MHC molecules in different orientations determining perhaps the geometry of the TCR/MHC interaction (43). However, how general is the

3

implication of junctional regions in contacting the peptide antigen awaits further evidence with additional peptide/MHC complexes.

PRIMARY STRUCTURE OF HLA-DR-RESTRICTED TCR SPECIFIC FOR AN IMMUNOGENIC PEPTIDE OF TETANUS TOXIN

To obtain further insights into the molecular mechanisms of antigen and MHC recognition, we have investigated the HLA-DR-restricted response to a T cell epitope defined by a short synthetic peptide comprising residues 830-844 (QYIKANSKFIGITEL) of tetanus toxin (tt830-844) (44). This peptide is universally immunogenic since it is recognized in most tetanus toxoid (TT)-primed donors tested, irrespective of their HLA-DR phenotype (45) and appears to bind to different HLA-DR alleles in a similar orientation, as indicated by experiments using truncated and substituted peptides (45, 46). This conclusion is also supported by the fact that some (promiscuous) clones are able to recognize tt830-844 in the context of several (up to five) HLA-DR allelles (47 and below).

This system therefore should allow one to determine the effect of the restricting element polymorphism (limited to the β chain of the HLA-DR) on the structure of TCR specific for the same immunogenic peptide. Polymorphic residues may be directly involved in contacting the TCR (e.g., those residues pointing away from the binding groove) and/or may modify the conformation of the peptide (e.g., polymorphic residues in the β sheets or on the α helix but pointing inside the binding groove) and thus effect the interaction of the peptide with the TCR (3).

A large number of clones [58] analyzed in this study (15) were obtained by *in vitro* stimulation with tt830-844 as previously described (45) from three TT-immunized donors of distinct HLA-DR haplotype and were restricted to HLA-DR6cI (donors AL and BR), DR4w4 (donor BR), and DRw11.1 (donor GC). The clones responded to low doses of peptide (0.01-1 µg/ml) and most of them proliferated to tenatus toxin presented by autologous APC demonstrating that they are indeed tetanus specific.

As summarized in Fig. 1, 50 to 75 % of the anti-tt830-844 T-cell clones derived from three donors were found to use a particular Vβ region gene segment, Vβ2. Thus, this response appears to be dominated by Vβ2 independently of the HLA-DR allele presenting tt830-844 suggesting that this Vβ segment may interact with a common determinant shared by these complexes. Interestingly, 10 of 12 clones utilizing Vβ2 displayed promiscuous recognition since they were able to efficiently recognize tt830-844 presented by autologous and other, although related, HLA-DR allelles (e.g. some clones recognized the peptide presented by HLA-DRw6cI and DRw11.2 or by these two plus HLA-DRw11.1, see Fig. 2).

4

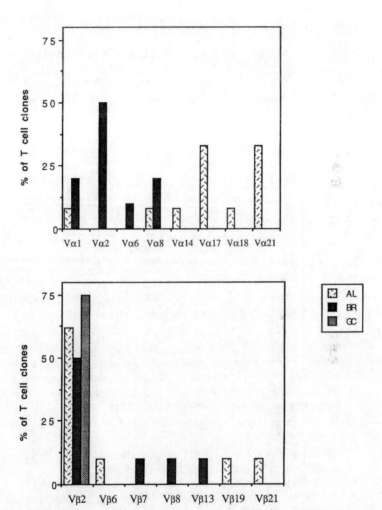

Figure 1. Vα and Vβ subfamilies used in anti-tt830-844 response in
three donors (AL, BR and GC).
Vβ2 usage was analyzed in the 3 donors (AL, BR and GC) by
using a proliferative assay of anti-tt830-844 T cell clones in
response to the TSST-1 (Toxic Shock Syndrome Toxin) which
is mitogenic for Vβ2[+] T cells (48). Vα and other Vβ subfamilies
were determined after anchor-PCR amplification and
sequencing of corresponding cDNA of 20 anti-tt830-844 T cell
clones from AL and BR donors.

CLONE T	HLA-DR RESTRICTION	Vα	GENE Jα	SEGMENTS[a] Vβ	Jβ
AL 15.3	DRw6cI P1	Vα21.1	JαR *	Vβ2.1a ◊	Jβ1.2
ALIII6.1	DRw6cI	Vα21.1	JαU *	Vβ2.1c	Jβ1.1
ALIII4.3	DRw6cI	Vα17.1	JαAF211 *	Vβ2.1	Jβ2.1
AL 17.3	DRw6cI	Vα17.1	JαF *	Vβ2.1c	Jβ2.7
AL 4.1	DRw6cI P1	Vα17.1 Vα14.2§	JαAC9 JαAA17	Vβ2.1a	Jβ1.5
AL 8.1 AL 12.1	DRw6cI P1	Vα8.1	JαIGRJa06	Vβ2.1c	Jβ2.7
AL 9.2	DRw6cI	Vα1.10	JαAA17	Vβ2.1c	Jβ2.1 °
BR 2.2	DRw6cI P2	ND		Vβ2.1c	Jβ1.2
BR 7.5	DRw6cI P3	Vα8.2 Vα2.6	JαK JαIGRJa10	Vβ2.1a	Jβ2.5
BR 9.13	DRw6cI P2	Vα2.6	JαR *	Vβ2.1c	Jβ2.5
BR 15.3	DRw6cI	Vα2.2	JαAF211	Vβ2.1c	Jβ2.7
BR 7.3	DR4w4	Vα1.2	JαAC17	Vβ2.1a	Jβ2.5
BR 22.5	DR4w4	Vα1.2	JαU	Vβ2.1c	Jβ2.4
BR 1.7	DR4w4	Vα8.1	JαS	Vβ2.1c	Jβ2.3

Figure 2. V and J gene segments used by the α and β chains of anti-tt830-844 TCR in clones from donors AL and BR.

a: previously described V and J gene segments were assigned according to nomenclature used in ref 15.

§: indicates a new V segment described in this analysis: Vα14.2 (15).

◊ Vβ2a and Vβ2c correspond to two alleles of Vβ2 differing by a single amino acid residue at position 10 (15).

P: indicates promiscuous clones which, in addition to autologous DR recognize tt830-844 presented in the context of DRw11.2 (P1), DRw11.1 and DRw11.2 (P2), or DRw11.1 (P3).

*: indicates clones in which a second, out-of-frame, sequence was detected.

°: a second in-frame Vβ sequence (Vβ17.1-Jβ2.6) has been found in 1 out of 9 independent sequences from this clone. ND: not done.

cDNA sequencing of 15 Vβ2-bearing TCR (all the clones utilize the Vβ2.1 gene segment) from two donors showed that they expressed 14 distinct junctional regions (Fig. 2 and 3) and that the Vα repertoire was rather heterogeneous. Thus, Vα21.1, Vα17.1, Vα2.2, Vα2.6, Vα8.1 and Vα1.10 germline gene segments could pair with Vβ2.1 to form receptors specific for the complex tt830-844/HLA-DRw6cI and, similarly, Vα1.2 and Vα8.1 formed receptors restricted to HLA-DR4w4. In few clones, two in frame Vα sequences were found (see Fig. 2 and 3), a consequence of a lack of allelic exclusion for the TCR α gene rearrangement (15, 40, 49). Although the specificity of these clones can not be unambiguously assigned to one of the two possible α/β pairs, it is likely that the Vα most frequently used in anti-tt830-843 response (e.g. Vα17.1 and Vα2) may be those pairing with Vβ2.1 to give the observed specificity. Transfection experiments have recently shown that this is indeed the case for clone AL4.1 whose specificity is born by Vα17.1 pairing with Vβ2.1 (unpublished results).

Interestingly, Vα21.1, Vα17.1, Vα2.2 and Vα8.1 display among themselves 50-60% amino acid sequence identity versus <40% with most other known members of the Vα family. This may indicate that these Vα segments have been co-selected together with Vβ2.1 for recognition of a similar peptide/MHC determinant.

However, as shown in Fig. 3, the junctional regions of α and β TCR specific, for instance, for tt830-843/HLA-DRw6cI are highly diverse both in amino acid composition and length, even when the same Vα and Vβ germline segments are utilized. Thus, no clear pattern of amino acid residue conservation in particular positions of the junctional regions of α and /or β could be identified as has been shown for other peptide/MHC complexes (33, 36, 37, 39). The only exception is represented by TCR of two clones (BR7.3 and BR22.5) both restricted to DR4w4 which use the same Vα (Vα1.2) and Vβ (Vβ2.1) germline segments and show a remarkable conservation (in both length and amino acid sequence) of the putative Vα CDR3. In this case, however, the Vβs CDR3 appear to be largely nonconserved.

Recently, two studies in the mouse investigating class I-restricted responses to defined peptides corresponding to choriomeningitis virus (39) and *Plasmodium berguei* (40) antigens found substantially different Vβ-D-Jβ junctional regions in the context of a common Vβ. Furthermore, similar to our findings, in one of these studies (40) the response was dominated by a particular Vβ segments which could pair with different Vα displaying also heterogenous junctional sequences. These and our results indicate that a rather composite repertoire of Vα/Vβ combinations may be selected for a response to some defined T-cell epitopes and that junctional regions may be totally, or in part nonconserved.

7

	Vα	Jα CDR3		Cα
AL15.3	Vα21.1-JαR	CAG	SYNARL MFGDGTQLVVKP	NIQ
ALIII6.1	Vα21.1-JαU	CAA	SGGQF YFGTGTSLTVIP	NIQ
ALIII4.3	Vα17.1-JαAF211	CAT	LYGONF VFGPGTRLSVLP	YIQ
AL17.3	Vα17.1-JαF	CAA	SSGNTPL VFGKGTRLSVIA	NIQ
AL4.1	Vα17.1-JαAC9	CAA	RQGGSEKL VFGKGTLTVNP	YIQ
	Vα14.2-JαAA17	CAS	RTGTASKL TFGTGTRLQVTL	DIQ
AL8.1	Vα8.1-JαIGRJa06	CAA	ENYGGSQGNL IFPKGTKLSVKG	NIQ
BR1.7	Vα8.1-JαS	CAA	EGPAGTAL IFGKGTTLSVSS	NIQ
BR15.3	Vα2.2-JαAF211	CAV	NMYDYGONF VFGPGTRLSVLP	YIQ
BR9.13	Vα2.6-JαR	CVV	NPNNARL MFGDGTQLVVKP	NIQ
BR7.5	Vα2.6-JαIGRJa10	CVV	AWKDM RFGAGTRPTVKP	NIQ
	Vα8.2-JαK	CAE	NSGGSNYKL TFGKGTLLTVNP	NIQ
AL9.2	Vα1.10-JαAA17	CVV	SADTGTASKL TFGTGTRLQVTL	DIQ
BR7.3	Vα1.2-JαAC17	CAV	SDGGAGNML TFGGGTRLMVKP	HIQ
BR22.5	Vα1.2-JαU	CAV	SDPGAGNOF YFGTGTSLTVIP	NIQ
BR2.2	ND			

	Vβ		Jβ CDR3		Cβ
AL15.3	Vβ2a	-Jβ1.2	CSA	LPRGYYGY TFGSGTRLTVV	EDL
ALIII6.1	Vβ2c	-Jβ1.1	CSA	RSDPATSA FFGQGTRLTVV	EDL
ALIII4.3	Vβ2	-Jβ2.1	CSA	RDPGGQAGFYNEQ FFGPGTRLTVL	EDL
AL17.3	Vβ2c	-Jβ2.7	CSA	RASPTYEQ YFGPGTRLTVT	EDL
AL4.1	Vβ2a	-Jβ1.5	CSA	RGGGRPQ HFGDGTRLSIL	EDL
AL8.1	Vβ2c	-Jβ2.7	CSA	KTGTSRYEQ YFGPGTRLTVT	EDL
BR1.7	Vβ2c	-Jβ2.3	CSA	PRGSGLTDTQ YFGPGTRLTVL	EDL
BR15.3	Vβ2c	-Jβ2.7	CSA	RDDSGLARASGTSYEQ YFGPGTRLTVT	EDL
BR9.13	Vβ2c	-Jβ2.5	CSA	LGLNQETQ YFGPGTRLLVL	EDL
BR7.5	Vβ2a	-Jβ2.5	CSA	SLPGLAIEETQ YFGPGTRLLVL	EDL
AL9.2	Vβ2c	-Jβ2.1	CSA	RGLPGTSGVSSYNEQ FFGPGTRLTVL	EDL
BR7.3	Vβ2a	-Jβ2.5	CSA	SSGIEGETQ YFGPGTRLLVL	EDL
BR22.5	Vβ2c	-Jβ2.4	CSA	TSGDKNIQ YFGAGTRLSVL	EDL
BR2.2	Vβ2c	-Jβ1.2	CSA	RGNYGY TFGSGTRLTVV	EDL

Figure 3. Amino acid sequence alignment of α and β V-(D)-J regions. For each clone, the names of the corresponding Vα, Jα, Vβ and Jβ segments are indicated. Only the last three amino acid residues of each V segment are shown, followed by the junctional sequences and the first three or four residues of the constant region. Amino acid residues are indicated using single-letter code. The assigment of the CDR3 loop is according to Chothia et al. (18). J sequences contributing to each CDR3 are underlined.

One interpretation for the apparent lack of coselection of specific junctional sequences in the context of a common V (Vβ and/or Vα) is that the junctional regions of these TCR do not play a major role in recognition and so are not structurally constrained. Evidence that putative CDR3 may not be critical for antigen recognition has been reported in the case of an I-A-restricted anti-arsonate clone (50). In this context, an interesting parallel with the structural constraints imposed by antigen on antibody selection is worth considering. Lack of particular sequence selection in CDR not involved in contacting antigen in antibody/antigen complexes of known three-dimensional structure, has been documented (51, 52).

That, however, at least for some TCR analyzed in our study, the putative CDR3 have an influence on recognition is shown by the effect of changes in these regions on DR restriction. Thus, clones AL 8.1 and BR1.7 (Fig. 2 and 3) expressing the same Vα and Vβ germline gene segments with different junctional sequences recognize tt830-844 in the context of DRw6cI and DR4w4, respectively. An effect on the HLA-DR restriction pattern (promiscuous recognition) due to changes in the putative CDR3 only, is also evident when comparing clone AL 15.3 which is able to see equally well HLA-DRw6cI, and DR11.2 (15 and our unpublished results) and ALIII6.1 which is exclusively restricted to HLA-DRw6cI. A similar example is represented by the TCR expressed by clones AL4.1 (promiscuous clone using Vα17.1 (unpublished results) and Vβ2.1) and clones AL17.3 or ALIII4.3 (see Fig. 2 and 3). A likely conclusion is that changes in the structure of the junctional regions are responsible for changes in the restriction (or restriction pattern) of the T-cell clones. According to the proposed model of MHC class II (3), amino acid differences between these alleles are located both in the β sheets and on the α helix of the DR β chain (Fig. 4) and may influence the conformation of the peptide or the direct interaction with the TCR. Therefore, further studies will be necessary to assess which one of these changes (in the peptide, the MHC or both) are responsible for selection of distinct junctional regions in these TCR.

Changes in the junctional regions of TCR recognizing the same antigen/MHC complex can be due to differences in fine specificity for the peptide, a hypothesis that can be tested by the use of substituted analogs of the peptide. Preliminary data examining the effect of multiple amino acid substitutions at each position (or truncation of the last residue, L) of tt830-844 on recognition by four Vβ2.1-expressing TCR restricted to HLA-DRw6cI (clones AL15.3, AL4.1, AL8.1 and AL17.3) indicate that the peptide segment seen by these TCR does not include the first (Q) and the last four (ITEL) residues. Rather, recognition is affected by substitutions in the central region of the peptide indicating that these TCR interact with the same stretch of the peptide.

Location of amino acids differing between:
DRw6cI and DR4w4

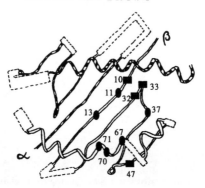

DRw6cI and DRw11.1   DRw6cI and DRw11.2

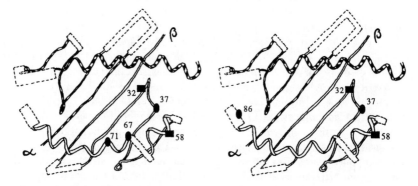

Figure 4. Comparison of the positions of polymorphic residues among
HLA-DR alleles presenting tt830-844 peptide.
Schematic representation of the predicted three-
dimensional structure of class II molecules according to
Brown et al (3). Polymorphic residues have been predicted
to point towards (circles) or outside (square) the antigen
binding site (3). The α chain is marked with stripes and
the β chain is in white.

Experiments are underway to examine which residues present in this
region are critical for binding to the HLA-DR and which ones interact
with the TCR.
    Regardless of possible differences in fine specificity, the strong
selection of Vβ2, but not of any particular junctional sequences, found
in many anti-tt830-844 clones, suggests that the former may be
responsible for most of the key contacts with this peptide/MHC
complex. As shown in Fig. 4, the HLA-DR alleles (plus the peptide)

selecting Vβ2 share a common (non-polymorphic) HLA-DR α chain but display 3 to 11 amino acid substitutions in the DR β chain (when compared to one another) located in positions that, according to the predicted MHC class II model (3), may influence the peptide conformation or may be in direct contact with the TCR. As can be seen in Fig. 4, these amino acid changes are restricted to one side of the pocket. Since Vβ2 is selected independently of these changes in the DR, it is tempting to speculate that Vβ2 would be rather interacting with the other side of the peptide-binding pocket contacting the DR molecule or the peptide or both.

Finally, an attractive hypothesis, in line with the proposed model of TCR recognition of peptide/MHC (5, 42) which would explain the CDR3 heterogeneity of TCR recognizing the same complex is that some particular peptides bound to MHC possess a degree of freedom allowing some of the peptide side chains to assume different conformations. This may result in creating different complexes selecting the same V regions contacting the MHC but very different junctions.

CONCLUDING REMARKS

The data summarized in the previous section indicate that recognition of a defined peptide/MHC complex, such as tt830-844/HLA-DR, does not necessarily imply selection of structurally homogeneous TCR variable regions. Thus, in spite of a strong selection for a particular Vβ germline segment, TCR specific for this complex can accomodate several Vα germline segments and, in most cases, entirely different junctional regions in both α and β chains.

When considering TCR specific for the same peptide/MHC complex, it is possible that different Vα segments (pairing with Vβ2) can be substituted with one another either because they share a structural similarity at positions (e.g. in the putative CDR) assuring key contacts or because these Vα do not contact the complex or contribute poorly to the overall interaction. In some instances, changes in the junctional regions, in the context of identical Vα and Vβ segments, could be correlated with differencies in MHC restriction patterns. It remains to be determined whether drastic changes in the junctional regions of anti-tt830-844 TCR displaying the same MHC restriction are due to alterations in fine specificity for the peptide or the MHC.

In view of the above results and of well documented X-ray crystallographic studies of antigen/antibody complexes in which the antibody heterodimer shows asymmetrical positioning with respect to antigen (53, 54), it remains entirely possible that not all the putative CDR of the TCR are simultaneously involved in recognition of a given complex. Furthermore, these regions may have interchangeable roles in contacting the peptide and the MHC.

Experiments using mono-substituted analogs of tt830-844 and reconstitution of TCR carrying mutations in the putative CDR should help to clearify whether our findings fit into the recently proposed model of the TCR/peptide/MHC interaction (5, 40) or if they reveal another example of a more complex strategy of TCR antigen recognition.

ACKNOWLEDGEMENTS

We wish to thank Dr. A. Smith for critical reading of the manuscript and C. Lefaucheux for help in editing. This work was supported by grants from the Pasteur Institute, Institut de la Santé et de la Recherche Médicale and Centre National de la Recherche Scientifique.

REFERENCES

1. L.A. Matis, The molecular basis of T-cell specificity, Annu. Rev. Immunol. 8:65 (1990).
2. P.J.Bjorkman, M.A. Saper, B. Samraoui, W.S. Bennett, J.L. Strominger, and D.C. Wiley, Structure of the human class I histocompatibility antigen, HLA-A2, Nature, 329:506 (1987).
3. J.H. Brown, T. Jardetzky, M.A. Saper, B. Samraoui, P.J. Björkman, and D.C. Wiley, A hypothetical model of the foreign antigen binding site of class II histocompatibility molecules, Nature, 332:845 (1988).
4. J.B. Rothbard, and M.L. Gefter, Interactions between immunogenic peptides and MHC proteins, Ann. Rev. Immunol. 9:527 (1991).
5. M.M. Davis, and P.J. Björkman, T-cell antigen receptor genes and T-cell recognition, Nature, 334:395 (1988).
6. B.A. Irving, and A.Weiss, The cytoplasmic domain of the T cell receptor ζ chain is sufficient to couple to receptor-associated signal transduction pathways, Cell, 64: 891 (1991).
7. A-M.K. Wegener, F. Letourneur, A. Hoeveler, T. Brocker, F. Luton, and B. Malissen, The T cell receptor/CD3 complex is composed of at least two autonomous transduction modules, Cell, 68: (1992).
8. R.K. Wilson, E. Lai, P. Concannon, R.K. Barth, and L. Hood, Structure, organization and polymorphism of murine and human T-cell receptor α and β chain gene families, Immunol. Rev, 101:149 (1988).

9. N. Kimura, B. Toyonaga, Y. Yoshikai, R.P. Du, and T.W. Mak, Sequences and repertoire of the human T cell receptor α and β chain variable region genes in thymocytes, Eur. J. Immunol., 17:375 (1987).

10. E. Jouvin-Marche, I. Hue, P. N. Marche, C. Liebe-Gris, J. P. Marolleau, B. Malissen, P. A. Cazenave, and M. Malissen, Genomic organization of the mouse T cell receptor Vα family, EMBO J., 9:2141 (1990)

11. A. Six, E. Jouvin-Marche, D. Y. Loh, P. A. Cazenave, and P. Marche, Identification of a T cell receptor β chain variable region, Vβ20, that is differentially expressed in various strains of mice, J. Exp. Med., 174:1263 (1991).

12. M.A. Robinson, The human T cell receptor β-chain gene complex contains at least 57 variable gene segments, J. Immunol., 146:4392 (1991).

13. B. Toyonaga, Y. Yoshikai, V. Vadasz, B. Chin, and T.W. Mak, Organization and sequences of the diversity, joining, and constant region genes of the human T-cell receptor β chain, Proc. Natl. Acad. Sci. US,. 82:8624 (1985).

14. S. Roman-Roman, L. Ferradini, J. Azocar, C. Genevee, T. Hercend, and F. Triebel, Studies on the human T cell receptor α/β variable region genes, Identification of 7 additional Vα subfamilies and Jα gene segments, Eur. J. Immunol., 21:927 (1991).

15. B. Boitel, M. Ermonval, P. Panina-Bordignon, R.A. Mariuzza, A. Lanzavecchia, and O. Acuto, Preferential usage and lack of junctional sequence conservation among human T cell receptors specific for a tetanus toxin-derived peptide: evidence for a dominant role of a germline-encoded V region in antigen/MHC recognition, J. Exp. Med., 175:765 (1992).

16. K. Ikuta, T. Ogura, A. Shimizu, and T. Honjo, Low frequency of somatic mutation in β-chain variable region genes of human T-cell receptor, Proc. Natl. Acad. Sci. USA., 82:7701 (1985).

17. J. Novotny, S. Tonegawa, H. Saito, D.M. Kranz, and H.M. Eisen, Secondary, tertiary, quaternary structure of T-cell-specific immunoglobulin-like polypeptide chains, Proc. Natl. Acad. Sci. USA., 83:742 (1986).

18. C. Chothia, D.R. Boswell, and A.M. Lesk, The outline structure of the T-cell αβ receptor, Embo J., 7:3745 (1988).

19. D.M. Becker, P. Patten, Y.H. Chien, T. Yokota, Z. Eshhar, M. Giedlin, N.R.J. Gascoigne, C. Goodnow, R. Wolf, K-I. Arai, and M.M. Davis, Variability and repertoire size of T-cell receptor Vα gene segments, Nature, 317:430 (1985).

20. L. Bourgueleret, and J.M. Claverie, Variability analysis of the human and mouse T-cell receptor β chain, Immunogenetics, 26:304 (1987).

21. R.D Jores, P.M. Alzari, and T. Meo, Resolution of hypervariable regions in T-cell receptor β chains by a modified Wu-Kabat index

of amino acid diversity, <u>Proc. Natl. Acad. Sci. USA.,</u> 87:9138 (1990).

22. P.J. Fink, L.A. Matis, D.L. McElligott, M. Bookman, and S.M. Hedrick, Correlations between T-cell specificity and the structure of the antigen receptor. <u>Nature,</u> 321:219 (1986).

23. A. Winoto, J.L. Urban, N.C. Lan, J. Goverman, L. Hood, and D. Hansburg, Predominant use of a Vα gene segment in mouse T-cell receptors for cytochrome c, <u>Nature,</u> 324:679 (1986).

24. M-Z. Lai, Y-J. Jang, L-K. Chen, and M.L. Gefter, Restricted V-(D)-J junctional regions in the T cell response to lambda expressor, <u>J. Immunol.,</u> 144:4851 (1990).

25. J.S. Danska, A.M. Livingstone, V. Paragas, T. Ishihara, and C.G. Fathman, The presumptive CDR3 regions of both T cell receptor α and β chains determine T cell specificity for myoglobin peptides, <u>J. Exp. Med.,</u> 172:27 (1990).

26. J. Wither, J. Pawling, L. Phillips, T. Delovitch, and N. Hozumi, Amino acid residues in the T cell receptor CDR3 determine the antigenic reactivity patterns of insulin-reactive hybridomas, <u>J. Immunol.,</u> 146:3513 (1991).

27. S.B. Sorger, Y. Paterson, P.J. Fink, and S.M. Hedrick, T cell receptor junctional regions and the MHC molecule affect the recognition of antigenic peptides by T cell clones, <u>J. Immunol.,</u> 144: 1127 (1990).

28. I. Engel, and S.M. Hedrick, Site-directed mutations in the VDJ junctional region of a T cell receptor β chain cause changes in antigenic peptide recognition, <u>Cell,</u> 54:473 (1988).

29. E.A. Nalefski, J.G. Wong, and A. Rao, Amino acid substitutions in the first complementarity determining region of a murine T-cell receptor chain affects major histocompatibility complex recognition, <u>J. Biol. Chem.,</u> 265: 8842 (1990).

30. C. Grégoire, N. Rebaï, F. Schweisguth, A. Necker, G. Mazza, N. Auphan, A. Millward, A.M. Schmidt-Verhulst and B. Malissen, Engineered secreted T-cell receptor αβ heterodimers, <u>Proc. Natl. Acad. Sci. USA.,</u> 88:8077 (1991).

31. Y. Lin, B. Devaux, A. Green, C. Sagerstöm, J.F. Elliott and M.M. Davis, Expression of T cell antigen receptor heterodimers in a lipid-linked form, <u>Science,</u> 249:677 (1990).

32. M.L. Silver, K.C. Parker, and D.C. Wiley, Reconstitution of MHC-restricted peptide of HLA-A2 heavy chain with β2-microglobulin in vitro, <u>Nature,</u> 350: 619 (1991).

33. J.L. Jorgensen, U. Esser, B. Fazekas de St.Groth, P.A. Reay, and M.M. Davis, Mapping T-cell receptor-peptide contacts by variant peptide immunization of single transgenics, <u>Nature,</u> 355: 224 (1992).

34. P.A. Morel, A.M. Livingstone, and C.G. Fathman, Correlation of T cell receptor Vβ gene family with MHC restriction, <u>J. Exp. Med.,</u> 166: 583 (1987).

35. M-Z. Lai, S.Y. Huang, T.J. Briner, J.G. Guillet, J.A. Smith, and M.L. Gefter, T cell receptor gene usage in the response to lambda repressor cI protein. An apparent bias in the usage of a V alpha gene element, J. Exp. Med., 168:1081 (1988).

36. H. Acha-Orbea, D.J. Mitchell, L. Timmermann, D.C. Wraith, G.S. Tausch, M.K. Waldor, S.S. Zamvil, H.O. McDevitt, and L. Steinman, Limited heterogeneity of T cell receptors from lymphocytes mediating autoimmune encephalomyelitis allows specific immune intervention, Cell, 54:263 (1988).

37. S.M. Hedrick, I. Engel, D.L. McElligott, P.J. Fink, M.L. Hsu, D. Hansburg, and L.A. Matis, Selection of amino acid sequences in the beta chains of the T cell antigen receptor, Science, 239:1541 (1988).

38. T. Aebischer, S. Oehen, and H. Hengardner, Preferential usage of $V\alpha 4$ and $V\beta 10$ T cell receptor genes by lymphocytic choriomeningitis virus glycoprotein-specific $H-2D^b$-restricted cytotoxic T cells, Eur. J. Immunol., 20:523 (1990).

39. D. Brändle, K. Bürki, V.A. Wallace, U.Hoffman Rohrer, T.W. Mak, B. Malissen, H. Hengartner, and H. Pircher, Involvement of both receptor Va and Vβ variable region domains and α chain junctional region in viral antigen recognition, Eur. J. Immunol., 21:2195 (1991).

40. J-L. Casanova, P. Romero, C. Widmann, P. Kourilsky, and J.L. Maryanski, T cell receptor genes in a series of class I major histocompatibility complex-restricted cytotoxic T lymphocyte clones specific for a *Plasmodium berghei* nonapeptide: implications for T cell allelic exclusion and antigen-specific repertoire,J. Exp. Med., 174.1371 (1991).

41. A.H. Taylor, A.M. Haberman, W. Gerhard, and A.J. Caton, Structure-function relationships among highly diverse T cells that recognize a determinant from influenza virus hemagglutinin, J. Exp. Med, 172:1643 (1990).

42. J.M. Claverie, A. Prochnicka-Chalufour, L. Bougueleret, Immunological implications of a Fab-like structure of the T-cell receptor, Immunol. Today., 8:202 (1988).

43. A. Prochnicka-Chalufour, L-L. Casanova, S. Avrameas, J-M. Claverie, and P. Kourilsky, Biased amino acid distribution in regions on the T cell receptor and MHC molecules potentially involved in their association, Int. Immunol., 3:853 (1991).

44. S. Demotz, A. Lanzavecchia, U. Eisel, H. Niemann, C. Widmann, and G. Corradin, Delineation of several DR-restricted tetanus toxin T cell epitopes, J. Immunol., 142:394 (1989).

45. P. Panina-Bordignon, A. Tan, A. Termijtelen, S.Demotz, G. Corradin, and A. Lanzavecchia, Universally immunogenic T cell epitopes: promiscuous binding to human MHC class II and promiscuous recognition by T cells, Eur. J. Immunol., 9:2237 (1989).

46. D. O'Sullivan, T. Arrhenius, J. Sidney, M.F. del Guercio, M. Albertson, M. Wall, S. Southwood, S. M. Colon, F. C. A. Gaeta, and A. Sette, On

the interaction of promiscuous antigenic peptides with different DR alleles. The identification of common structural motifs, J. Immunol., 147:2663 (1991).

47. R.W. Karr, P. Panina-Bordignon, W-Y. Yu, and A. Lanzavecchia, Antigen-specific T cells with monogamous or promiscuous restriction patterns are sensitive to different HLA-DRβ chain substitutions, J. Immunol., 4242:4247 (1991).

48. Y. Choi, J.A. Lafferty, J.R. Clements, J.K. Todd, E.W. Gelfand, J. Kappler, P. Marrack, and B.L. Kotzin, Selective expansion of T cells, J. Exp. Med., 172:981 (1990).

49. M. Malissen, J. Trucy, F. Letourneur, N. Rebaï, D.E. Dunn, F.W. Fitch, L. Hood, and B. Malissen, A T cell clone expresses two T cell receptor a genes but uses one αβ heterodimer for allorecognition and self MHC-restricted antigen recognition, Cell, 55:49 (1988).

50. K-N. Tan, B.M. Datlof, J.A. Gilmore, A.C. Kronman, J.H. Lee, A.M. Maxam, and A. Rao, The T cell receptor Vα3 gene segment is associated with reactivity to p-azobenzenearsonat, Cell., 54:247 (1988).

51. E.A. Padlan, G.H. Cohen, and D.R. Davis, On the specificity of antibody/antigen interactions: phosphorylcholine binding to McPC603 and the correlation of three-dimensional structure and sequence data, Ann. Inst. Pasteur/Immunol., 136C:271 (1985).

52. P.M. Alzari, S. Spinelli, R.A. Mariuzza, G. Boulot, R.J. Poljak, J.M. Jarvis, and C. Milstein, Three-dimensional structure determination of an anti-2-phenyloxazolone antibody: the role of somatic mutation and heavy:light chain pairing in the maturation of an immune response, EMBO J., 9:3807 (1990).

53. R. Tulip, J.N. Varghese, R.G. Webster, G.M. Air, W.G. Laver, and P.M. Colman, Crystal structures of neuraminidase-antibody complexes, Cold Spring Harbor Symposia on Quantitative Biology, Volume LIV, Cold Spring Harbor Laboratory Press (1989).

54. A.G. Amit, R.A. Mariuzza, S.E.V. Phillips, and R.J. Poljak, The three-dimensional structure of an antigen-antibody complex at 2.8 Å resolution, Science, 233:747 (1986).

2

STRUCTURE OF THE TCR-Ag-MHC COMPLEX

Nadine Gervois, Bing-Yuan Wei, Paolo Dellabona,
Jean Peccoud, Christophe Benoist, Diane Mathis

Laboratoire de Génétique Moléculaire des Eucaryotes du
CNRS et U/184 INSERM, Institut de Chimie Bilogique
Faculté de Médecine, Strasbourg, France

INTRODUCTION

T cells recognize foreign antigen via a tri-molecular complex composed of the T cell receptor (TCR), an antigenic peptide (Ag), and a molecule encoded by the major histocompatibility complex (MHC). Much effort has been devoted to solving the crystal structure of the TCR-Ag-MHC complex, but this formidable feat has so far been unsuccessful. In the meantime, several models of the interacting triad have been proposed (1-3) on the basis of the marked sequence homology between TCRs and immunoglobulins and on the known structure of MHC class I molecules surmised from X-ray crystallographic studies (4,5).

Attempts to evaluate these models have taken two perspectives - from the TCR's point of view and the MHC molecule's. We will discuss these in turn.

HOW THE T CELL RECEPTOR "SEES" ANTIGEN AND THE MHC MOLECULE

A major prediction of the models of Davis and Bjorkman (1) Chothia et al (2) and Claverie et al (3) was that the V(D)J junctional regions of the TCR alpha and beta chains would be responsible for recognition of antigen, while other less variable regions would contact the two MHC domains that form the antigen-binding groove. Various approaches have been taken to test this prediction:

Sequence correlations

Several groups have determined the sequences of TCR alpha and/or beta chains from sets of T cell clones or hybridomas specific for a particular antigen in the context of a particular MHC molecule (6-16) . The hope was to find a correlation between the sequences of variable stretches of the TCR and Ag or MHC usage.

Such a correlation was certainly found in some cases. Perhaps the most striking example was the case of T cells specific for a peptide of myelin basic protein in the context of A^u (7,8). These cells expressed receptors with strongly biased V_α and V_β usage and of quite restricted sequence in the V(D)J junctional, or CDR3 regions. Other studies found correlations between Ag + MHC specificity and V_α (eg 9) or V_β (eg 11) usage, or both (eg 12) - sometimes with restricted CDR3 sequences, sometimes without. Still other studies (often unpublished) found no feature of the TCR that correlated with Ag + MHC specificity.

It is difficult to draw an overall conclusion from these data. In some cases there is support for the proposed models of the TCR-Ag-MHC complex, but in many cases there is not. However, one should be aware that in some (most?) of the experiments it is not completely sure that the Ag used, even if a peptide, carried only one T cell epitope.

CDR3 swaps

The approach here was to take two T cell clones with different, but usually related, Ag specificities: X and Y; to isolate their TCR α and β genes; to make chimeric constructs by exchanging particular segments of the X and Y TCRs; and finally, to transfect these constructs into the appropriate T cells to determine whether the antigen specificities have been converted from one to the other. Several attempts have been made to "swap" Ag specificities by "swapping" the CDR3 regions of the alpha chain, the beta chain, or both. None have so far been successful (M. Davis, personal communication; N. Glaichenhaus, personal communication). Thus, this approach has failed to provide support for the proposed models of the trimolecular complex.

Transgenic mouse experiments

Two groups have exploited TCR transgenic mice to probe the relationship between Ag specificity and the structure of the T cell receptor. The strategy here was a) to begin with a transgenic mouse that carries an already rearranged α or β gene derived from a clone of particular Ag specificity, b) to inject Ag into the half-receptor transgenic mouse, and c) to determine by sequence analysis the structure of the other TCR chain in clones that come up in response to the Ag. This strategy can be rendered even more powerful by injecting analogues of the original Ag known to be altered at TCR interaction residues; "escape" T cell clones that come up are likely to have alterations at T cell receptor sites that directly contact antigen.

Brändle et al (17) used transgenics carrying rearranged TCR α or β genes from an H-2D^b restricted clone specific for residue 32-42 of the lymphocyte choriomeningitis virus glycoprotein. With the β-trangenic mice, they found that after, but not before, Ag injection there was a tremendous bias for TCRs that bore V 2, the same variable region carried by the original clone. In addition, there was a striking restriction of Jα usage. With the α-transgenic mice, they found after Ag injection, and again not before, a bias for TCRs that bore either $V_\beta 8.1.$ or $V_\beta 8.3$; the original clone carried $V_\beta 8.1$. The beta chain CDR3 regions were really quite diverse, although there was somewhat preferential usage of one particular Jβ.

Jorgensen et al (18) employed mice carrying rearranged TCR α and β genes from a T cell clone that was specific for moth cytochrome c peptide 88-103 in the context of E^k. When they injected the wild-type peptide, which has a positively charged residue at position 99, into the β-transgenics, essentially all of the α chains on T cells which came up carried the $V_\alpha 11.1$ variable region, like the original clone. Interestingly, one position in the CDR3 region was always occupied by a negatively charged residue. When, on the other hand, they injected a mutant peptide, which had a negatively charged residue at position 99, essentially all of the α chains on responding T cells carried $V_\alpha 11.1$, but the same position in the CDR3 segment was now occupied by a positively charged residue. Results somewhat parallel were obtained with the α-transgenics and peptides differing at position 102; namely that a charge change in the peptide provoked a charge change in a CDR3 residue. However, in this case there was a striking degeneration of the specificity of variable segment usage - several different V_βs were used, none of which corresponded to that employed by the original clone.

Taken together, these results on half-TCR transgenics are both supportive and non-supportive of the proposed models of the TCR-Ag-MHC complex. Certainly they provide evidence that Ag contacts residues in the CDR3 segment of the TCR. But the strongly biased V_α and V_β usage and the induction of degenerate V_β usage by a single amino acid change in the peptide are somewhat difficult to reconcile with the models' predictions.

HOW THE MHC MOLECULE "SEES" ANTIGEN AND THE T CELL RECEPTOR

An alternate perspective is from the point of view of the MHC molecule: is the manner in which it contacts Ag and the TCR consistent with the models proposed by Davis and Bjorkman, Chothia et al and Claverie et al? The major approach to answering this question has been site-directed mutagenesis of particular MHC alleles, almost always confined to their polymorphic residues.

We will focus on our own studies of the murine class II molecule A^k (19-21). Our experiments had as an underlying assumption the molecular modelling of Brown et al (24). These workers exploited the HLA-A2 crystallographic structure to model the Ag binding site on MHC class II molecules. Their model predicts that peptides bind to class II molecules in a groove - one side formed by an α-helix from the alpha chain, the other side by an α-helix from the beta chain, and the floor by β-sheets from both subunits. The proposed similarity to HLA-A2 is supported by a number of arguments, including: 1) class I and II complexes have highly analogous functions in antigen presentation and T cell repertoire selection; 2) the same TCR variable regions can recognize both class I and class II molecules (eg 23); 3) identical peptides can be presented by both classes of MHC molecule (eg 24); 4) class I and II domains have highly similar spacings of conserved and polymorphic residues (22); and 5) low-resolution spectroscopic studies suggest comparable secondary structures (25).

With these considerations in mind, we have performed a detailed mutational analysis of the alpha chain of the A^k complex. Both a "horizontal" and a "vertical" series of mutations have been created and extensively characterized.

"Horizontal" substitutions

In the first set of mutations, residues 50-79 of the alpha chain were changed one by one to an alanine, and the effect of each mutation on the stimulation of T cells determined. We concentrated on these positions because they include the two most prominent allelically hypervariable regions (53-59 and 69-77) and because they have been predicted to form one face of the antigen-binding groove. We mutated both conserved and polymorphic residues, unlike all previous studies. Alanine was the substitution of choice for several reasons: it has the simplest side-chain with chirality; ala replacements generally provoke only minor disturbances in secondary structure; and ala is a common replacement for all but the aromatic amino acids, according to sequence comparisons between evolutionarily related proteins.

The substitutions were made via oligonucleotide-directed mutagenesis of a cDNA carried in an expression vector. Each mutant A_α^k cDNA was transfected into L cells together with a wild-type A_β^k cDNA also in an expression vector, and a selection marker. Cells expressing high levels of A^k were isolated by cytofluorimetric sorting and then expanded. By this means, we created a panel of 30 mutant antigen presenting cell (APC) lines.

We measured with some care the ability of these APC lines to present a hen egg lysozyme peptide (HEL 46-61) to a set of T cell hybridomas. Fully one third of the mutations had essentially no effect, provoking no more than a five-fold reduction in presentation efficiency to any of the hybridomas. Others were somewhat more consequential, measurably reducing the efficiency to one or two of the hybridomas, but not significantly affecting presentation to the remainder. Finally, some of the mutations had drastic effects, virtually abolishing presentation to at least three of the hybridomas (≥ 250 fold down) and greatly reducing the efficiency to the fourth. The most drastic changes were those at positions 56, 62, 65, 66, 68 and 69.

We have transposed these data onto the model of Brown et al and found a most striking pattern. The critical residues lined up with the periodicity of the α-helix and all faced either up toward the TCR or into the peptide groove. The most interesting finding was that for different T hybridomas, even those specific for the same peptide, there was marked heterogeneity in the precise pattern of residues critical for a response. That there was heterogeneity in the critical up-pointing residues was not really very surprising; one would expect that different hybridomas expressing different TCRs could make variable MHC contacts either at the top of the helix or on its side. That there was heterogeneity in the critical in-pointing residues was much more surprising. The extreme comparison is of kLy11.10 with kLy4.10: recognition by the former was affected by mutations at in-pointing residues at every turn of the helix from position 56 to position 72; recognition by the latter was affected by only two in-pointing residues, 56 and 66.

It was entirely possible that the heterogeneity in recognition pattern displayed by the four HEL-specific T hybridomas was peculiar to the HEL46-61 peptide. It was important to evaluate the ability of our mutant APC lines to present a second peptide, and so we evaluated presentation of RNase 41-61 to two hybridomas. Heterogenous recognition patterns were again detected: only two residues were critical for presentation to one hybridoma, several more for presentation to the other; the most devastating Ala replacement for recognition by one hybridoma had little effect on recognition by the other, and vice versa.

These surprising results forced us to question whether the in-pointing residues predicted by the model of Brown et al were indeed antigen contact sites. The antigen presentation data with the various hybridomas suggested a means to answer this based on competition experiments. We noticed that the residues critical for presentation of RNase 41-61 to the hybridoma, TS12, were largely non-overlapping with those crucial for presentation of HEL 46-61 to the different hybridomas. Thus, we incubated TS12 with various APC lines in the presence of a sub-saturating amount of the ribonuclease peptide and increasing concentrations of the lysozyme peptide. We expected that APC lines capable of binding HEL 46-61 would show a progressively reduced capacity to present RNase 41-61, while lines incapable of binding the competitor would always exhibit the same presentation ability. To our astonishment, the competition curves for all of the mutant lines were much like the curves for the wild-type APC line, implying that they did not show any significant binding impairment.

Studies on peptide analogues seemed to offer an alternative strategy for confirming which A_α^k residues serve as antigen contacts. The inability of wild-type A^k molecules to present a particular peptide variant could potentially be reversed by a compensatory mutation in the class II molecule; alternatively, the inability of a particular variant A^k molecule to present wild-type peptide might be reversed by mutation of the peptide. A large panel of RNase 43-56 analogues was employed to this end. Thirty analogues were screened on the wild-type and mutant APC lines; all unexpected presentations to the RNase-specific hybridoma were retested more extensively. In the end, we could identify one A mutation that could compensate for non-presentation of variant RNase peptides. Wild-type A^k molecules could present the wild-type RNase 43-56 peptide, containing a Leu at position 51, but not the analogues containing Val or Ile at this position; the mutant complex with an Ala substitution at Aα position 66 could present all three of the peptides. These results were particularly interesting because RNase amino acid 51 has previously been identified as a contact residue for the MHC molecule (26).

By these and related studies using truncated peptides (20) or influenza virus variants (27), we could confirm that several of the in-pointing residues delineated by Brown et al are indeed antigen contact residues. Thus we were faced with the paradoxical finding that no single residue seems required for binding of an antigenic peptide. Different hybridomas specific for the same peptides seem to depend on different contacts being made between the MHC molecule and Ag.

"Vertical substitutions"

One possible criticism of the above experiments is that the mutations were always substitutions of alanine- a small, neutral amino acid which might not cause very drastic perturbations. To counter this cristicism, we created a second set of 11 mutants: in each case the thr at position 69 was replaced by a different amino acid. These replacements were chemically diverse, including positively (Arg) and negatively (Glu) charged as well as bulky (Trp, Tyr) residues. The mutant cDNAs were again expressed in L cells together with a wild-type A_β^k cDNA, and the mutant APC lines that were derived were tested for their ability to present HEL 46-61 and RNase 41-61 to a panel of T cell hybridomas.

The results were really quite striking. All of the mutants were recognized by at least one hybridoma specific for HEL 46-61, indicating that this peptide could bind to all of them. This included the Trp replacement, which is so bulky that one would have predicted it might block off the groove. Yet different hybridomas had vastly different sensitivities to the various replacements, ranging from 3A11 which accepted none of the substitutions to 2B5.1 which accepted them all.

We are left in the same quandry as with the set of "horizontal" replacements: it seems that different T cells specific for the same peptide rely on different contacts being made between Ag and the MHC molecule. A similar dilemma presented itself to Krieger et al in their studies on HLA-DR7 mutations (28) and Brett et al (29) in their analysis of myoglobin peptide analogues. At present , we see two ways to side-step this dilemma:

First, it might be that one must take a more sophisticated view of the TCR-Ag-MHC triplex. There may not be such a clear distinction between TCR-Ag, Ag-MHC and MHC-TCR contacts as previously believed. It is possible that the TCR can "reach" into the groove and interact with MHC residues that are also contacting Ag; different TCRs could easily "reach" to different extents. It is also possible that the MHC molecule moulds the antigen into a particular conformation and different contours of this moulded structure are recognized by different T cells.

Second, it might be that a single peptide can bind differently within the groove when participating in different TCR-Ag-MHC complexes. In the absence of a T cell of appropriate specificity, it would fit loosely in the groove, resonating between various positions or conformations while making multiple, non essential contacts. When the T cell arrives, its receptor would "dock on" and "lock in" a particular disposition.

CONCLUSION

The sum of data from both perspectives does not provide convincing support for models of the TCR-Ag-MHC complex proposed by Davis and Bjorkman, Chothia et al and Claverie et al. As models often are, they appear far too simple. It seems that regions other than the V(D)J junction might contact antigen and that the V(D)J junction might contact both antigen and the MHC molecule. It seems that different TCRs might interact differently with the same Ag presented by the same MHC molecule.

Many are confident that X-ray crystallography of a TCR-Ag-MHC molecule will resolve the question definitively. But we should be aware that evéen X-ray crystallography has its limitations: it will provide us with a static structure and will need to be performed on several complexes composed of the same Ag and MHC molecule but different TCRs - a daunting task.

REFERENCES

1. Davis, M.M., and Bjorkman, P.J. (1988) Nature 334,395-402.
2. Chothia, C., Boswell, D.R., and Lesk, A.M. (1988) EMBO J; 7,3745-3755.
3. Claverie, J.M., Prochnicka-Chalufour, A., and Bougueleret, L. (1989) Immunol. Today, 10,10-13.
4. Bjorkman, P.J., Saper, M.A., Samraoui, B., Bennett, W.S., Srominger, J.L., and Wiley, D.C. (1987a) Nature 329,506-512.

5. Bjorkman, P.J., Saper, M.A., Samraoui, B., Bennett, W.S., Strominger, J.L., and Wiley, D.C. (1987b) Nature, 329,512-518.

6. Sorger, S.B., Hedrick, S.M., Fink, P.J., Bookman, M.A., and Matis. L.A. (1987) J. Exp. Med. 165,279-301.

7. Acha-Orbea, H., Mitchell, D.J., Timmermann, L., Wraith, D.C., Tausch, G.S., Waldor, M.K., Zamvil, S.C., McDevitt, H.O., and Steinman, L. (1988) Cell 54,263-273.

8. Urban, J.L., Kumar, V., Kono, D.H., Gomez, C., Horvath, S.J., Clayton, J., Ando, D.G., Sercarz, E.E., and Hood, L. (1988) Cell 54,577-592.

9. Tan, K-N., Datlof, B.M., Gilmore, J.A., Kronman, A.C., Lee, J.H., Maxam, A.M., and Rao, A. (1988) Cell 54,247-261.

10. Lai, M-Z., Jang, Y-J., Chen, L-K., and Gefter, M.L. (1990) J. Immunol. 144,4851-4856.

11. Danska, J.S., Livingstone, A.M., Paragas, V., Ishihara, T., and Fathman, C.G. (1990) J. Exp. Med.172,27-33.

12. Aebischer, T., Oehen, S., and Hengartner, H. (1990) Eur. J. Immunol. 20,523-531.

13. Yanagi, Y., Maekawa, R., Cook, T., Kanagawa, O., Michael, B., Oldstone, A. (1990). J. Virology 64,5919-5926.

14. Koseki, H., Imai, K., Ichikawa, T., Hayata, I., and Taniguchi, M. (1989). Internat. Immunol. 1,557-564.

15. Wither, J., Pawling, J., Phillips, L., Delovitch, T., and Hozumi, N. (1991) J. Immunol. 146,3513-3522.

16. Kempkes, M., Palmer, E., Martin, S., von Bonin, A., Eichmann, K., Ortmann, B., and Weltzien, H.U. (1991) J. Immunol. 147,2467-2473.

17. Brändle, D., Bürki, K., Wallace, V.A., Hoffmann Rohrer, U., Mak, T.W., Malissen, B., Hengartner, H., and Pircher, H. (1991) Eur. J. Immunol. 21,2195-2202.

18. Jorgensen, J.L., Esser, U., Fazekas de Groth, B., Reay, P.A., and Davis, M.M. (1991) Nature, in press.

19. Dellabona, P., Peccoud, J., Kappler, J., Marrack, P., Benoist, C., Mathis, D. (1990) Cell 62,1115-1121.

20. Peccoud, J., Dellabona, P., Allen, P., Benoist, C., and Mathis, D. (1990) EMBO J. 9,4215-4223.

21. Wei, B-W., Gervois, N., Mer, G., Adorini, L., Benoist, C., Mathis, D. (1991) Int. Immunol. 3,833-837.

22. Brown, J.H., Jardetzky, T., Saper, M.A., Samraoui, B., Bjorkman, P.J., and Wiley, D.C. (1988) Nature 332,845-850.

23. Rupp, F., Acha-Orbea, H., Hengartner, H., Zinkernagel, R., and Joho, R. (1985) Nature 315,425-427.

24. Perkins, D.L., Lai, Ming-Zing, Smith, J.A., and Gefter, M.L. (1989) J. Exp. Med. 170,279-283.

25. Gorga, J.C., Dong, A., Manning, M.C., Woody, R.W., Caughey, W.S., and Strominger, J.L. (1989) Proc. Natl. Acad. Sci USA 86,2321-2325.

26. Lorenz, R.G., Tyler, A.N., and Allen, P.M. (1989). J. Exp. Med. 170,203-207.

27. Warren, A.P., Paschedag, I., Benoist, C., Peccoud, J., Mathis, D. and Thomas, D.B. (1990) EMBO J. 9,3849-3856.

28. Brett, S.J., McKean, D., York-Jolley, J., and Berzofsky, J.A. (1989) Int. Immunol. 1,130-140

29. Krieger, J.I., Karr, R.W., Grey, H.M., Yu, W-Y., O'Sullivan, D., Batovsky, L., Zheng, Z-L., Colon, S.M., Gaeta, F.C.A., Sidney, J., Albertson, M., del Guercio, M-F., Chesnut, R.W., and Sette, A. (1991) J. Immunol. 146,2331-2340.

3

IS RECOGNITION BY THE T CELL RECEPTOR HIGHLY CONSTRAINED OR VERY DEGENERATE?

Navreet K. Nanda and Eli E. Sercarz

Department of Microbiology & Molecular Genetics
University of California at Los Angeles
Los Angeles, CA

Evidence is accumulating that the paradox implied in the title represents the essential characteristic of T cell recognition. Like all other receptor-ligand systems, T cell receptor molecules show constraints in interaction with their ligands, an essential attribute underlying the specificity of antigen recognition. We have attempted to define the limits of these constraints and have shown that for recognition of some antigen-determinants, there is a strict requirement for usage of specific V_β gene segments as constituents of T cell receptors. In contrast, we present evidence that individual T cell clones seem able to recognize multiple structures, an evidence for flexibility in receptor-ligand interaction. We discuss these two contrasting attributes of T cell recognition in the context of T cell ontogeny and models of T cell receptor-ligand interaction.

Positive selection of T cell repertoire in the thymus: evidence for inherent flexibility in ligand recognition displayed by individual TCR molecules

T cells recognize antigen only when their determinants are associated with MHC molecules, an attribute called MHC-restriction. Those T cells that bear the receptors binding to self-MHC within an "appropriate" affinity range are selected for further development. Only the positively selected cells, after undergoing subsequent negative selection to remove cells with high affinity for self-peptides, are allowed to become mature T cells, constituting the peripheral immune system.

T Lymphocytes: Structure, Function, Choices, Edited by F. Celada
and B. Pernis, Plenum Press, New York, 1992

The generation of mature T cells requires an interaction of TCRs expressed on immature T cells with MHC molecules present in the thymus as has been shown by two lines of evidence: anti-class II antibodies specifically interfere with the differentiation of class-II restricted cells; more dramatically, in mice transgenic for rearranged TCR $\alpha\beta$ chains, transgene $\alpha\beta$TCR positive thymic precursors mature into T cells only if mice also express the MHC ligand specific for the transgenic TCR molecules[1]. The TCR transgenic mice yielded unequivocal proof for the commonly held concept that selection of T cells in the thymus takes place in the absence of nominal antigen[1].

This model of imprinting of self-MHC restriction, if strictly applied, yields a paradox that has been at the center of numerous debates: the ligand recognized by the T cells undergoing selection is a self-MHC molecule, complexed to self-peptides, whereas the specific ligand recognized later by the same T cells in the periphery is either (a) self-MHC plus foreign peptides, or (b) in alloreactive situations, foreign MHC plus self-peptide. Differential "affinity" of TCRs has been at the center of most arguments presented to explain the above paradox: it has been proposed, for example, that TCRs have an inherent affinity for MHC molecules, and that the selective steps mentioned earlier lead to the recognition of self-MHC plus self-peptides with low affinity, but self-MHC coupled with foreign peptides with high affinity[2]. Irrespective of the relative affinities, it would seem clear that individual T cell receptor molecules must have the ability to interact with at least two ligands: (i) the thymic ligand self-peptide plus a self-MHC molecule, as well as (ii) the peripheral ligand- foreign peptide plus self-MHC. However, at the level of fine-specificity, the TCR displays an apparently highly specific ligand-receptor recognition, showing (a) strict requirements for TCR V gene usage for a given ligand[3-7] and vice-versa, (b) strict requirement for specific sequences of amino acids in the ligand for a given TCR molecule - implying additional attributes and constraints that have previously remained undiscussed.

The potential diversity of the TCR repertoire : the expected plasticity

The T cell repertoire has an almost unlimited potential available for diversification by rearrangement of V, D and J gene segments, and introduction of junctional residues: somewhere between 10^{12} to 10^{18} unique heterodimeric structures can be constellated[8]. Four inbred strains and numerous wild mouse strains have recently been shown to have lost half of the V_β gene segments from their germ-line repertoire[9]. These mice define a new genotype at the V_β locus- the TCR V_β truncated (V_β^a) genotype. Given such a huge potential for T cell repertoire diversification, the absence of 10 TCR V_β gene segments should not have impinged on the capacity of the remaining half of the TCR V_β gene segments to develop specificities to any particular ligand.

EVIDENCE FOR CONSTRAINTS IN TCR RECOGNITION

Experimental test of plasticity of the TCR repertoire: unexpected limitations of plasticity of the repertoire as well as constraints in T cell recognition

The question of plasticity of TCR repertoire became more interesting as well as testable when it was found that in response to certain antigen determinants, T cell clones use a very limited number of TCR V_β gene segments[3]. This observation, combined with the existence of mice missing a large number (50%) of TCR V_β gene segments (V_β^a), led us to examine those antigenic determinants, the response to which was limited to gene segments absent in V_β^a mice: (i) sperm-whale myoglobin 111-121/I-E^d: all T cell clones specific for this determinant in DBA/2 mice were shown to express TCR V_β 8.2[5] and (ii) myelin basic protein 1-11/I-A^u: all T cells of this specificity show restricted usage of TCR V_β 8.2 and V_β13 gene segments in B10.PL mice[4]. Fortuitously for our studies, both the V_β 8.2 and V_β 13 gene segments are absent from V_β^a haplotype mice. As the TCR V_β^a genotype is expressed in nature in only a few laboratory strains, but not in the H-2^d and H-2^u haplotypes, needed for studying the SWM 111-121/ I-E^d and MBP 1-11/I-A^u responses, alternatives had to be devised for this study. Therefore, responses to myoglobin were investigated in two different V_β^a, H-2^d strains of mice: (a) recombinant inbred (RI) strains of mice, (CxJ)3 and (CxJ)8, and (b) the congenic strain, B10.D2.βL. To study H-2^u restricted responses to myelin basic protein (MBP), radiation bone marrow (BM) chimeras of the type: V_β^a (C57L, H-2^b)-> V_β^b (C57BL/6 x B10.PL)F1, H-2^{bxu}) were constructed.

Four striking observations reflecting constrained recognition of p110-121/I-E^d, were made when BALB/c and B10.D2 mice were compared with (CXJ)3 and B10.D2$_\beta$L for their ability to produce T cells specific for SWM p110-121:

(i) Complete absence of a T cell proliferative response to p111-121/I-E^d in lymph node cells of mice lacking TCR V_β 8 gene segments: an unusual result of immunization of BALB/c mice with p110-121 was that this peptide could induce T cells with two different specificities, one restricted by I-A^d and the other restricted by I-E^d. As we primarily aimed to investigate the plasticity of response to the I-E^d- restricted determinant, which was reported to show exclusive usage of the TCR V_β 8.2 gene segment, p111-121 along with p110-121 were employed in our experiments as the former was only able to induce E^d-restricted T cells, and could therefore be used to probe for I-E^d-restricted responses[3]. (The I-E^d-restricted response to this determinant has subsequently been shown, in fact to be restricted to the hybrid I-A^d/I-E^d (I-A^d/I-E^d) MHC molecule[3, 10].

BALB/c (V_β^b-H-2^d) and (CxJ)3 (V_β^a, H-2^d) mice, primed with SWM p110-121 respond very strongly to added native SWM. The (CxJ)3 mice show a considerably reduced response to peptide 110-121 compared to BALB/c mice. However, the most interesting observation was that (CxJ)3 mice with their truncated repertoire, could raise no response to p111-121, the peptide which is able to induce only hybrid I-A^d/I-E^d-restricted T cells. Analogous results, showing a lack of any response to p111-121, were obtained in B10.D2βL (V_β^a) mice. Meanwhile, a strong T cell response to p111-121 was obtained in the BALB/c and B10.D2 (V_β^b) strains of mice (Table 1).

(ii) Absence of p110-121-specific, I-A^d/I-E^d-restricted T hybrids cloned from T cell lines obtained from V_β^a mice: V_β^a mice, despite the presence of the necessary I-A^d, I-E^d and hybrid restriction elements did not show any response to the 111-121 peptide even in long-term T cell lines specific for p110-121. All hybrids obtained from the (CxJ)3 T cell line responded to peptide 110-121 in the context of the I-A^d molecule: no hybrid could be found with the hybrid A^d/E^d restricted specificity[3].

Table 1

EVIDENCE FOR LACK OF PLASTICITY OF TCR REPERTOIRE

Response to SWM p110-121 in primed lymph nodes or cloned T hybrids

	V_β^b (wild-type) mice	V_β^a (TCR truncated) mice
Restricting molecule	BALB/c, B10.D2	[CXJ]3, B10.D2βL
I-A$_\alpha^d$/I-E$_\beta^d$	+++	-
I-A^d	++	++

(iii) Lack of plasticity in I-A^d-restricted specificity: It appeared that the I-A^d-restricted response to p110-121 was less constrained than the I-A^d/I-E^d-hybrid-restricted response, as both V$_\beta$^{a} and V$_\beta$^{b} mice could respond to this specificity. However, when we examined the two separate panels of p110-121-specific, I-A^d-restricted T cell hybrids derived from V$_\beta$^{a} or V$_\beta$^{b} mice for fine specificity, we were surprised to learn that distinct determinants within the same peptide 110-121 were being recognized by these two panels of T cells. The determinant recognized by V$_\beta$^{a} T cells is N-terminal (core: 110-118) with an absolute requirement for the residue ala-110 for a successful interaction with TCRs. On the other hand, V$_\beta$^{b} T cells focus on a C-terminal region (core: 112-118) on the same peptide with an absolute requirement for C-terminal residue 118[11]. Thus, V$_\beta$^{a} mice are not able to recognize the two most dominant determinants: (i) 111-121, restricted by hybrid I-A^d/E^d and (ii) the C terminal determinant of p110-121, restricted by A^d. They instead respond with T cells specific for a third, distinctly N-terminal determinant[11]. This evidence suggests that at the level of individual determinants within a mutideterminant peptide, serious constraints in recognition exist at the level of V$_\beta$ usage.

(iv) Lack of response to MBP p1-11/I-A^u in chimeric mice reconstituted with V$_\beta$^{a} bone marrow: When chimeric mice of the types C57BL/6 (V$_\beta$^{b})- > (C57BL/6xB10.PL)F1, and C57L (V$_\beta$^{a})- > (C57BL/6xB10.PL)F1 were compared for their ability to respond to p1-11 of MBP, the former chimeras with a V$_\beta$^{a} TCR repertoire, gave a normal T cell proliferative response to this peptide, whereas the latter chimeras reconstituted with V$_\beta$^{a} T cells, showed no response to this peptide. These results demonstrate the absolute requirement for TCR V$_\beta$ 8.2 or 13 in the mouse repertoire in order to make a detectable response to MBP 1-11[3].

EVIDENCE FOR PLASTICITY IN RECOGNITION BY INDIVIDUAL T CELLS

Surprisingly, and in sharp contrast to the constraints observed in recognition of the ligands, p110-121/I-E^d and MBP1-11/I-A^u, we found evidence for recognition of multiple ligands by individual, cloned T cell hybrids recognizing these same two determinant regions: (1) SWM 110-121/I-A^d and (2) MBP 1-11/I-A^u.

(1) Degeneracy in requirements of TCR recognition in SWM 110-121/I-A^d-restricted T hybrids:

According to the premise described in the first section, we decided to search for degeneracy in recognition displayed by individual TCR molecules, using the strategy described below. We made use of the well-characterized panel of I-A^d-restricted, SWM 110-121-specific T cell hybridomas. The minimal core of the determinant recognized by this panel of T hybrids is 112-118 and residues 117 and 118 of SWM 110-121 are essential for triggering the T cell receptors of these T cells. We truncated this peptide determinant so as to remove crucial residues required for TCR contact, resulting in a non-stimulatory peptide. Thus, peptide 105-116 is a non-stimulatory fragment for the SWM 110-121-specific cells, even though we showed that it could bind to the I-A^d molecule. In the hope of creating alternative antigenic conformations that might activate one or more of the T cells in our panel, we made amino acid substitutions within the peptide 105-116[12].

Interestingly, changes at each of 4 positions of 105-116 generated a peptide structure that could trigger different T hybridomas in the absence of residues 117 and 118. The data are summarized below and in Table 2.

The changes that converted a negative fragment into a stimulatory peptide were:
(i) a conservative change at position glu-109 to asp-107: a conservative substitution at position 109 (glu to asp) resulted in a peptide analog that was now highly stimulatory for 2 of 6 I-A^d-restricted T cell hybridomas. Position 109 is outside the core of the I-A^d-restricted determinant[12].
(ii) a conservative change at position ile-112 to val-112: the substituted p105-116, val-112 is stimulatory for the CM8-6 T hybrid while 105-116, leu-112 is non-stimulatory.
(iii) a conservative change at position val-114: the substituted peptide, p105-116, leu-114 as well as p105-116, ile-114 are strongly stimulatory for the T cell hybrid CJM4-16[12].
(iv) dual conservative changes at position glu-109 and ile-112: in preliminary experiments, an analog with two conservative substitutions, at positions 109 (glu to asp) and 112 (ile to leu), was another very strong stimulator for a set of three of the six I-A^d-restricted T cell hybrids. Single substitutions at either of these two positions were not stimulatory for these hybrids.

We conclude that residues important for triggering the TCR of an individual T cell hybridoma in the substituted (truncated-core)-peptide 105-116, are different from from native (complete-core)-peptides, such as 110-121 or 112-123. Thus, receptor molecule can be triggered by peptides with apparently different structures.

(2) Recognition of dual ligands by individual MBP 1-9-specific T cell clones: Specific recognition of distinct unrelated ligands by individual T cell clones could also be shown in the MBP system. On immunization of B10.PL and (SJL x B10.PL) F1 mice with MBP, the majority of the response is directed towards the N-terminal, I-A^u-restricted acetylated Ac1-9/Ac1-11, while 35-47, an I-E^u-restricted determinant, is subdominant[13]. Each of these determinants can cause EAE. The original aim in our study was to define the mechanisms underlying the dominance of Ac1-11. For this purpose, we synthesized a chimeric peptide (CP) in which Ac1-11 was joined directly to 35-47 with a peptide bond. The CP was found to be much better in recalling the primed response to the chimeric immunogen than to its separate components[13]. While defining the dominant chimeric determinants, presumably bridging the two peptides, it was found that many of the Ac1-9-specific long term lines, clones and hybridomas, could be activated by a set of chimeric peptides, including CP7-11:35-47[14]. None of these clones could be activated directly by 35-47 or by 9-11:35-47. Previous studies had mapped out the MHC and TCR contact residues in Ac1-9 as the N-acetyl group and residue 4 as important for MHC binding. Since the evidence of cross-reactivity in our system was surprising, it was important to define the residues critical for recognition of Ac1-9. In fact Ac1-6 itself was stimulatory, but only at 10-fold excess of the peptide while no reactivity was obtained to AcAAAAAASQR. However, the CP's had an entirely different requirement : extensive amino acid replacement within 7-11 region showed that residues 10 and 11 (SK) were the only essential residues. Of

Table 2

DEGENERACY IN REQUIREMENTS FOR RECOGNITION BY SWM 110/121/1A[d] SPECIFIC T HYBRIDS.

IL-2 RESPONSES

Hybrids:	CM8-16	CM8-18	CM8-30	CM8-34	CM8-6	CJM4-16
Peptides						
110-121	+++	+++	+++	+++	+++	+++
105-118	+++	+++	+++	+++	+++	+++
105-116	-	-	-	-	-	-
Substituents of p105-116:						
glu-109 to asp-109	-	-	-	++	++	-
Ile-112 to val-112	-	-	-	-	++	-
val-114 to leu-114	-	-	-	-	-	++

course, residues from 35-47 within the CPs are completely unrelated to Acl-6 (14 and Table 3). Apparently, this represents a situation of "shape-mimicry" in which en-bloc, an alternative peptide cassette fits into the MHC groove and exposes an epitypic terrain for recognition that simulates a variety of Acl-9-specific clones.

Table 3

Sequences of peptides Acl-9 and chimeric peptide 7-11:35-47

Acl-9:   A S Q K R P S Q R

Ac7-11:35-47:   S Q R S K : I L D S I G R F F S G D R

---

(3) Alloreactivity and degeneracy: Early work by Sredni and Schwartz[15] demonstrated that T cell clones recognizing peptides from foreign antigens in an MHC context also could recognize certain allogenic targets. Conversely, it has recently been demonstrated that specific peptides within the allo-MHC context are recognized by particular T cell clones: some DRW11-specific human alloreactive T cell require a defined peptide of human serum albumin for allorecognition[16]. Transfection of a single $\alpha\beta$ TCR molecule could impart reactivity to foreign antigen plus self-MHC simultaneously with allo-recognition in two reported cases[17, 18], showing that allorecognition represents inherently redundant recognition. In an analogous situation, H.-K Deng in this laboratory (unpublished), has defined two distinct ligands recognized by a single T cell clone: one peptide (nominal antigen) recognized in the context of self-MHC (I-A^d) another recognized in the context of an allo-MHC molecule (I-A^k). These observations together would indicate that a single TCR molecule is capable of recognizing multiple ligands.

Our studies described above provide evidence for degeneracy of T cell receptor recognition. Furthermore, they essentially indicate that the redundant recognition is not limited to odd antigens or systems, but is clearly widespread and a fundamental characteristic of immune recognition.

CONCLUSIONS

What is the significance of the two diametrically opposed attributes characteristic of TCR recognition?

Let us first consider the constraints observed in the TCR V_β gene usage by a given antigen determinant:
(i) The non-plasticity of the TCR repertoire as evidenced above would have important implications for proposed models of TCR recognition of its ligand, the peptide-MHC complex. According to the Davis and Bjorkman model of TCR structure, the V-regions of TCR α and β chains form the first and second hypervariable regions of the

molecule which only interact with non-polymorphic regions of the MHC molecule[8]. The third hypervariable region, formed by the junctional region between V, (D) and J gene segments of the two chains, is largely responsible for the nominal antigen-specificity of the TCR molecules. This model, if strictly applied, would fail to explain our results that the presence or absence of particular V_β chains strongly influences the determinant recognized by T cell receptor structures, as multiple (if not all) V chains should be able to interact with the MHC (I-A^d) molecules. Restriction of TCR V_β gene usage by T cells recognizing a given antigen has been reported frequently[3-7]. In a most recent example, murine T cells specific for a nonapeptide of <u>Plasmodium berghei</u> show expression of the TCR V_β 13 gene segment in 60% of cells in spite of diversity in their $V\alpha$ chain as well as the joining (CDR3) regions of both the α and β chains[6]. Even more surprising is the restricted usage of the TCR V_β 2 gene segment reported in human T cells specific for a tetanus toxoid peptide, despite the utilization of a variety of MHC molecules as restriction elements[7]. The results presented here, along with those recently reported by Casanova et al.[6] and Boitel et al.[7] would argue for the idea that the V_β segment of the TCR molecule plays a significant role in establishing key contacts with the peptide-MHC complex as a whole. In fact, the results of Boitel et al. describing restricted TCR V_β gene usage, even among T cells recognizing the same peptide and restricted to different Class-II molecules suggest that the peptide acts as a selective force for choosing the V_β segment used by T cells.

(ii) Our results would also have implications for responsiveness of an individual to an antigen as a whole. It is now well known that both MHC genes and non-MHC (self-superantigen) genes provide a potent influence (the former by both positive and negative selection and the latter by negative selection) to regulate the expression of TCR V_β gene segments in the periphery[19,20]. If the response to a dominant peptide is restricted to one or two V_β gene segments, the lack of these segments might result in loss of the response to the whole antigen as well as to the determinant in question. However, if the determinant region were responded to in three ways, for example, and two genes were missing in the strain, a residual response might appear, which in fact would be a highly limited one.

Now, we may turn to the plasticity or degeneracy of recognition at the level of individual TCR molecules.

(i) Degenerate TCR recognition is a required implication of the concept of thymic positive selection in the absence of nominal antigen. It would be difficult to methodically seek out examples of degenerate recognition, although the approach we attempted with the myoglobin peptide was successful. The level of degeneracy, i.e. the number of different MHC and or peptide ligands that could substitute for each other may never be fully known. Nevertheless, whatever the actual mechanism of TCR recognition, it is apparent that there is substantial evidence for degeneracy and redundancy at the level of single T cell clones, a requirement for models of positive selection in the thymus. Positive selection is an efficiency requirement in the peripheral repertoire: T cells must recognize antigens strictly in the context of those MHC molecules expressed in the individual.

(ii) The immune system has been designed to successfully confront any challenge with pathogens; meanwhile, clever pathogens are driven to devise structural alternatives in their struggle to outwit the immune system and survive. In this host versus pathogen struggle, some inherent redundancy would be helpful to the host, enabling it to easily respond to minor changes in the pathogen antigens.

(iii) A major disadvantage of redundant recognition is the feature that a determinant on a foreign peptide in an MHC context is similar to a self peptide-self MHC ligand. Plasticity in TCR recognition would proportionally increase the chances of activation of self-reactive T cells by look-alike foreign antigens, resulting in autoimmunity.

Evidently, during evolution, a balance was reached between these two extremes - constraints and plasticity in T cell receptor recognition - resulting in a moderately, but not infinitely flexible immune system.

REFERENCES

1.   von Boehmer, H., 1990, Development biology of T cells in T cell transgenic mice. Ann. Rev. Immunol. 8:531.
2.   Janeway, C.A. Jr., Carding, S., Jones, B., Murray, J., Portoles, P., Rasmussen, R., Rojo, J., Saizawa, K., West, J. and Bottomly, K., 1988, CD4+ T cells: specificity and function. Immunol. Rev. 101: 39.
3.   Nanda, N.K., Apple, R. and E. Sercarz., 1991, Limitations in plasticity of the T cell receptor repertoire. Proc. Natl. Acad. Sci. 88:9503.
4.   Urban, J.L., Kumar,V., Kono D., Gomez, C., Horvath S.J., Clayton J., Ando, D.J., Sercarz, E.E., and Hood, L.L., 1988, Restricted use of T cell receptor V genes in murine autoimmune encephalomyelitis raises possibilities for antibody therapy. Cell 54: 577.
5.   Morel, P.A., Livingstone, A.M., and Fathman, C.G., 1987, Relation of T cell receptor V_β gene family with MHC restriction. J. Exp. Med. 166: 583.
6.   Casanova, J-L., Romero, P., Widmann, C., Kourilsky, P., and Maryanski, J.L., 1991, T cell receptor genes in a series of class-I MHC-restricted T lymphocyte clones specific for a Plasmodium berghei nonapeptide: implications for T cell allelic exclusion and antigen-specific repertoire. J. Exp. Med. 174: 1371.
7.   Boitel, B., Ermonval, M., Panina-Bordignon, P., Mariuzza, R.A., Lanzavecchia, A. and Acuto, O., 1992, Preferential V_β gene usage and lack of junctional sequence conservation among human T-cell receptors specific for a tetanus toxoid-derived peptide: evidence for a dominant role of a germ-line-encoded V region in antigen/MHC recognition. J. Exp. Med. 175: 765-777
8.   Davis, M.M. and Bjorkman, P.J., 1988, T cell antigen receptor genes and T cell recognition. Nature 334:395-402.
9.   Behlke, M.A., Chou, H.S., Huppi, K. and Loh, D., 1986, Murine T-cell receptor mutants with deletions of β-chain variable region genes. Proc. Natl. Acad. Sci. 83, 767-771.
10.   Ruberti, G., Gaur, A., Fathman, C.G., and Livingstone, A.M., 1991, The T cell receptor repertoire influences V_β element usage in response to myoglobin. J. Exp. Med. 174:83.
11.   Nanda, N. K., Arzoo, K., and Sercarz, E. E., 1992, In a small multi-determinant peptide, each determinant is recognized by a different V_β gene segment. J. Exp. Med. In Press.
12.   Nanda, N. K., Geysen, M. and Sercarz, E.E., 1992, Degeneracy in requirements of peptide structures for T cell receptor recognition. Manuscript in preparation.

13. Bhardwaj, V., Kumar, V.,Geysen, H.M. and Sercarz, E.E., 1992, Subjugation of dominant immunogenic determinants within a chimeric peptide.   Eur. J. Immunol. In press.

14. Bhardwaj, V., Kumar, V., and Sercarz, E.E., 1992, Molecular mimicry in the absence of protein sequence homology. In preperation.

15. Sredni, B. and Schwartz, R. H., 1980, Allorectivity of an antigen-specific T cell clone. Nature 287: 857.

16. Panina-Bordignon, P., Corradin, G. , Roosnek, E., Sette, A. and Lanzavecchia, A., 1991, Recognition by Class II alloreactive T cells of processed determinants from human serum proteins. Science 252: 1548.

17. Kaye, J. and Hedrick, S.M. ,1988, Analysis of specificity for antigen, Mls, and allogenic MHC by transfer of T-cell receptor alpha- and beta-chain genes. Nature 336: 580.

18. Portoles, P., Rojo, J.M., Janeway, C.A. Jr., 1989, Assymetry in recognition of antigen: self-class-II MHC and non-self-class-II MHC molecules by the same T cell receptor. J. Mol. Cell Immunol. 4:129.

19. Woodland, D. L., Happ, M.P., Gollob, K.J., and Palmer, E., 1991, An endogenous retrovirus mediating deletion of $\alpha\beta$ cells? Nature 349: 529.

20. Dyson, P.J., Knight, A.M., Fairchild, S., Simpson, S. and Tomonari, K., 1991, Genes encoding ligands for deletion of $V_{\beta}11$ T cells cosegregate with mammary tumor virus genomes. Nature 349:531.

KEYWORDS: T cell receptor (TCR)/ TCR recognition/ TCR repertoire/ TCR V_{β} genes/ positive selection/plasticity of recognition/ degenerate recognition/ constrained recognition/ auto-immunity/ SWM/MBP/receptor-ligand interactions.

4

CHARACTERISTICS OF FOREIGN AND SELF PEPTIDES ENDOGENOUSLY BOUND TO MHC CLASS

I MOLECULES

Grada M. van Bleek and Stanley G. Nathenson

Albert Einstein College of Medicine
1300 Morris Park Avenue
Bronx, NY 10461

INTRODUCTION

The immune response against intracellular pathogens is mediated by cytotoxic T lymphocytes (CTL) which recognize components of the invader in combination with self molecules of an infected cell. These self products are encoded in the class I region of the Major Histocompatibility Complex (MHC) and the process is termed MHC restriction (1). Originally it was thought that proteins of the infectious agent were present in close proximity to the surface MHC molecule where they could be simultaneously contacted by the T cell receptor (TCR). Therefore it was hypothesized that spike proteins of viruses which are present on cell surfaces after fusion with the cell membrane at cell entry, or before virus budding, would be likely candidates for such a dual recognition. However, from studies with influenza viruses with variations in the nonsurface exposed nucleoprotein (Np) it was learned that, unlike antibodies, the major population of influenza specific CTL recognize the Np rather than the surface exposed hemagglutinin (2). Studies in which the Np gene or parts thereof were transfected into L-cells further showed that the influenza CTL were specifically recognizing a linear peptide stretch in Np (3). Moreover the specific CTL response could be triggered when short synthetic peptides were mixed in vitro with antigen presenting cells carrying the appropriate MHC class I molecules (4).

Recent crystallographic studies of human class I molecules HLA-A2 and HLA-AW68 have shed more light on the structural strategy for peptide recognition by revealing the configuration of the peptide-MHC complex (5,6,7). The two membrane distal outer domains of the class I molecules form a groove in which the antigenic peptide is bound. The crystallized molecules contained extra-electron dense material in the antigen binding groove that could not be accounted for by the MHC sequence, suggesting that MHC and the bound peptides had actually been co-crystalized. These peptides were most likely derived from cellular proteins and judging from the hydrophobic nature of the antigenic groove might play a role in stabilizing the MHC heavy chain.

After the early experiments which showed that it was possible for CTL to recognize target cells exposed to synthetic peptides in vitro, a search began for the antigenic peptides involved in virus specific CTL responses by using synthetic peptides. After several peptide epitopes were discovered, attempts were made to find common characteristics which would allow one to predict class I restricted viral peptides (8,9,10). These predictions were not always accurate and the question remained as to which factors most influenced the presentation of peptides by class I molecules: class I binding specificity, protein degradation processes, or the availability of peptides in the cellular compartments in which binding to the MHC molecules occurs. In order to address such questions it was clear that insight was needed into the characteristics of peptides that are actually presented by class I molecules such as MHC binding motifs, specific length requirements, and the presence of particular amino acid residues that might provide clues to the involvement of certain proteolytic enzymes.

Isolation of a naturally processed viral peptide recognized by Vesicular Stomatitis Virus (VSV) specific CTL

In order to study MHC bound peptides, we set up a procedure to isolate endogenously produced peptides presented by murine class I molecules H-2K,D, using Vesicular Stomatitis Virus infected EL4 cells (H-2^b) (11). Our strategy was to isolate the class I molecules from such cells and analyze them for the viral antigenic peptide which we would first determine by testing a set of synthetic peptides. In addition to searching for a peptide with a known sequence, the use of this viral system had the advantage that VSV shuts down cellular protein synthesis after the production of viral proteins begins. Because class I molecules were thought to be occupied with peptides from cellular proteins in the absence of an infection, this relative advantage of viral protein synthesis would result in a larger representation of the viral antigenic peptide(s) in the complete K^b bound peptide mix. We decided to further increase the sensitivity of our detection methods by using metabolic radiochemical labeling with ^{3}H amino acids, an approach which had the additional advantage that radiolabeled peptides that were recovered would be products of intracellular protein synthesis.

Because it was already known from previous studies that the K^b restricted CTL response against VSV was directed against nucleoprotein (N) (12) and most likely against a peptide in the N terminal one third of the molecule, we set out to define the peptide epitope, using pepscan methodology. We made a set of 75 overlapping peptides (13 amino acids long) that covered the amino terminal 161 amino acids of N. These peptides were tested in a ^{51}Chromium release assay for their ability to sensitize H-2K^b target cells for recognition by VSV specific CTL. We found that four overlapping peptides of the panel were recognized by the VSV specific CTL clone. These peptides had a common core of seven amino acid residues: GYVYQGL.

When the K^b presented antigenic peptide of VSV was defined, we radiolabeled VSV infected EL4 (H-2^b) cells at a time during which viral protein synthesis was high, class I synthesis still continued, yet total cellular protein synthesis was low. After immunoprecipitation of the K^b molecules from these cells, the bound peptide fractions were isolated. A comparison of the HPLC profile of peptides eluted from K^b molecules from infected cells with those of peptides similarly eluted from K^b products of uninfected cells showed one major extra peak in the profile of infected cells (Fig. 1). The peptide corresponding with this peak was eight amino acids long and had a pattern of radiolabeled amino acid residues which was consistent with the core of amino acids that was established with the synthetic peptide panel. This major virus peak (N52-59) was the only prominent peak in the K^b-peptide profiles of VSV infected cells when ^{3}H Tyr, ^{3}H Leu, ^{3}H Lys and ^{3}H Arg labels were used. To approach the question of whether the peptide was the only epitope seen in VSV infection, cold target blocking experiments were performed. We used a system measuring the response of VSV specific bulk CTL tested on H-2K^b positive target cells that were transfected with the complete N gene, in the presence of EL4 cells treated with the synthetic peptide corresponding to the natural sequence. We showed that the synthetic epitope was identical with the peptide presented by the cells transfected with the N gene. Thus the naturally processed peptide isolated and characterized by our procedure was the only major antigenic peptide provided to K^b by N protein. Because N protein is the only viral protein which contains a major K^b restricted CTL epitope, the endogenously produced peptide we detected was the only major antigenic peptide in the entire virus.

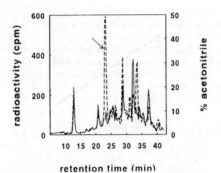

Fig. 1.   HPLC analysis of the radiolabelled peptide fractions of MHC class I K^b molecules. ^3H Arg and ^3H Leu were used as metabolic labels. Peptides from infected EL4 cells (—); peptides from uninfected EL4 cells (- - -); acetonitrile gradient (···). The virus specific peak with biological activity is indicated with an arrow.

The naturally produced antigenic peptide of VSV is bound to K^b by Tyr residues at positions three and five and the main chain atoms of the amino terminus

In order to obtain information on the three dimensional conformation of the natural VSV peptide when it is complexed to K^b for presentation to T cell receptors, we prepared single Ala substituted peptides of the natural octamer: N52-59 Arg-Gly-Tyr-Val-Tyr-Gln-Gly-Leu, and tested their ability to bind to K^b and trigger CTL responses (13). To measure their binding ability we used a peptide competition assay in which the Ala-analogs were used to compete with a K^b restricted nonapeptide of Sendai virus (SV) in a ^{51}Cr release assay with SV specific bulk CTL. To test their ability to sensitize H-2K^b positive target cells for CTL recognition, regular ^{51}Cr release assays were performed in which the analog peptides were tested on H-2K^b positive target cells with a panel of VSV specific cloned CTL. The analog peptides that were unable to compete with the SV nonamer peptide, allowed us to define which amino acid residues were important for K^b interaction. These were the Tyr residues at P (positions) 3 and P5 and the Leu residue at P8 in the peptide.

Replacement of the amino acid residues at positions that were not involved in MHC interaction all influenced the recognition of at least one CTL clone in the panel, indicating that the Ala substitution influenced T cell receptor contact, either by directly altering the interaction of that amino acid side chain with the TCR of the clone in which recognition was affected, or by influencing the three dimensional orientation of a neighboring TCR-contact amino acid residue. None of these amino acid residues of this category abrogated the recognition of all of the clones of the tested panel, indicating that the changes at position one, two, four, six and seven only influence TCR contact but not K^b interaction.

The length of the naturally produced VSV peptide was eight amino acids. This length also appeared to be optimal for binding to K^b since truncated peptides shortened at either the N- or C- terminus were extremely inefficient competitors in the SV competion assay. An interesting further observation was that Arg at position one in the peptide had a dual role. The side chain of Arg appeared to be involved in TCR contact because replacement with Ala affected the recognition of 4 out of 5 clones of a panel tested on target cells primed with the analog (13). The replacement with Ala did not affect binding of this analog to K^b. However when the amino acid residue at position 1 is absent, K^b binding was dramatically impaired, suggesting that main chain atoms of the first amino acid are implicated in K^b interaction. Thus in conclusion, for this particular K^b restricted antigenic peptide, the presence of amino acid residues Tyr at P3 and P5 and Leu at P8 as well as the amino terminus are important for K^b interaction. The amino acids at the other positions irregularly interspersed between the anchor residues all contribute to the specific three dimensional picture which is recognized by T cell receptors (Fig. 2).

↑ TCR CONTACT RESIDUES

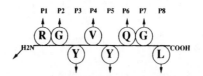

↓MHC CONTACT RESIDUES

Fig. 2.   Schematic representation of the orientation of the amino acid side chains in N52-59 when bound to H2-K^b. P refers to the position in the peptide.

Endogenous self peptides bound to K^b molecules have a major common motif of Tyr or Phe residues at positions three and five

We used the procedure we developed for the isolation of the viral peptide presented by K^b molecules of VSV infected cells, to isolate K^b bound peptides of cellular origin (14). Peptides isolated from the antigenic groove of K^b molecules that were immunoprecipitated from nonviral-infected Concanavalin-A stimulated spleen cells of C57BL/6 mice showed a frequent occurrance of Tyr or Phe amino acid residues at P3 and P5 relative to the amino terminus (Table 1). Because these residues are so frequently present

Table 1. Partial radiolabel sequence analysis of peptides eluted from K^b, K^bm1, K^bm8 and D^b molecules.

MHC	%ACN	1	2	3	4	5	6	7	8	9	10	
K^b	23.5	—	—	Y	—	Y	—	—	—	—	—	
	24.2	—	—	Y	—	—	—	—	—	—	—	
	25.0	—	—	Y	Y	Y	—	—	—	—	—	
	25.3	—	—	Y	—	—	—	—	—	—	—	
	26.4	—	—	—	—	Y	—	Y	Y	—	—	
	27.3	—	—	Y	—	—	—	Y	—	—	—	Mix?*
	30.8	—	—	—	—	—	—	—	—	—	—	
	35.9	—	—	Y	—	—	—	Y	—	—	—	
	46.0	—	—	—	—	—	—	Y	—	—	—	
	47.3	—	—	—	—	—	Y	Y	—	—	—	
K^bm1	23.1	—	—	—	—	—	—	—	—	—	—	
	23.8	—	—	—	—	Y	Y	—	—	—	—	
	27.4	—	—	—	—	—	—	—	—	—	—	
	28.6	—	—	—	—	Y	—	Y	—	—	—	
	35.0	—	—	—	—	—	Y	Y	—	—	—	
	46.1	—	—	—	—	—	—	Y	—	—	—	
	47.4	—	—	—	—	—	—	Y	—	—	—	
K^bm8	23.5	—	—	—	—	—	—	—	—	—	—	
	25.5	—	—	Y	Y	Y	—	—	—	—	—	
	27.5	—	—	Y	—	Y	—	—	—	—	—	
	28.5	—	—	Y	—	—	Y	—	—	—	—	Mix?
	30.9	—	—	Y	—	—	—	—	—	—	—	

MHC	%ACN	1	2	3	4	5	6	7	8	9	10	
D^b	23.1	—	—	—	—	—	—	—	—	—	—	
	23.5	—	—	—	—	—	—	—	—	—	—	
	25.9	—	—	—	—	—	—	—	—	—	—	
	26.2	Y	—	—	Y	—	—	Y	—	Y	—	
	26.3	—	—	Y	—	—	—	Y	Y	—	—	
	42.7	—	—	—	—	—	—	—	—	—	—	Mix?
K^b	26.3	F	—	—	—	—	—	—	—	—	—	
	27.0	—	—	F	—	F	—	—	—	—	—	
	32.7	—	—	F	F	F	—	—	—	—	—	
	33.0	F	—	—	—	—	—	—	—	—	—	
	33.5	—	—	—	—	F	—	—	—	—	—	
	35.6	—	—	F	—	F	—	—	—	—	—	
K^b	17.3	—	—	L	L	—	—	—	—	—	—	
	23.5	—	—	L	—	—	—	—	—	—	—	
	25.3	—	—	L	—	—	—	—	—	—	—	
	27.2	—	—	L	—	—	—	—	L	—	—	
	28.7	—	—	L	—	—	L	—	L	—	—	
K^bm8	32.0	—	—	—	—	—	—	—	—	—	—	
	33.9	—	—	—	—	—	—	—	—	—	—	Mix?
	17.3	—	—	—	L	—	—	—	—	—	—	
	30.5	—	—	—	L	—	—	—	—	—	—	

*When multiple signals of similar yield were found within the run of 10 cycles the amino acid residues were considered to be located in the same peptide. When they were of different yields the sample was considered to be a mixture and where this was likely it is indicated in the table as mix?

Single letter code for amino acids: Y, Tyr; F, Phe; L, Leu.

at those particular positions in the peptides and are virtually absent at other positions in the complete set of HPLC peak fractions after labeling with ^{3}H Tyr or ^{3}H Phe, this strongly suggests that, in general, similar to the isolated case of the VSV peptide, peptides bound to H-2K^b use phenyl side chains at P3 and P5 as anchors. Because we did not find these amino acid residues at positions two and four, or four and six, this also suggests that the positioning of the amino terminus relative to the anchor residues may be important for K^b interaction.

The three-phenyl-five-phenyl binding motif is specific for K^b and is not found in peptide sets bound to the other allelic product on H-2^b cells: D^b or the K^b mutant molecule K^{bml}

The binding motif of class I bound peptides is extremely specific for each individual MHC molecule (14). Three amino acid changes in the K^{bml} mutant molecule, two of which alter residues whose side chains point into the antigen binding groove, dramatically alter the set of peptides bound to this molecule as shown in Table 1. The three amino acid changes in K^{bml} are clustered in the α2-α helix and only locally affect the features of the antigen binding groove. D^b which differs from K^b in many more amino acids lining the antigen binding groove also shows dramatic changes in the sets of peptides bound. Therefore the three dimensional structure of the antigen binding groove and local changes in charge and/or hydrophobicity appear to be important factors determining which internally produced peptides are presented on the cell surface.

The rules that determine peptide presentation by MHC class I molecules

The experiments described above showed the common characteristics of peptides bound to one particular MHC molecule and the specificity of peptide selection imposed by the features of the antigen binding groove. These include the presence of a class I binding motif, a certain distance between the amino terminus and the first anchor residue and a restricted total length (this paper: the VSV antigenic peptide is an octamer and Falk et al. (15): peptides bound to specific MHC allelic products have a certain length). However the presence of a peptide binding motif in a certain protein sequence is not sufficient to ensure peptide presentation. For instance the nucleoprotein of VSV contains a second peptide with the H-2K^b binding motif, yet it is not found in the HPLC profiles of K^b peptides eluted from VSV infected cells. This observation suggests that other requirements have to be met before antigen presentation will occur. The absence of the second peptide containing the K^b binding motif might mean that this peptide is not produced with the right length, escapes transport into the E.R., or contains amino acid residues elsewhere in the peptide, which are incompatible with the three dimensional space available in the K^b groove.

Indications for the importance of the entire peptide sequence in determining a fit in the antigen binding groove are the discovery of antigenic peptides which do not conform to the length requirements for a particular MHC molecule. The Adenovirus 10mer peptide (16) and Sendai virus nonamer peptide (17) are both one amino acid longer than the consensus length of peptides bound to their presenting molecules respectively D^b and K^b (15). Interestingly these peptides both contain multiple proline residues which would introduce kinks in the peptide backbone and which would allow them therefore to still position the anchor residues in the proper orientations, despite the fact that they are longer.

In addition to the specific presence of amino acid residues within the antigenic sequence, amino acid residues adjacent to the optimal binding sequence may play a crucial role by facilitating the proteolytic processes that produce the MHC bound peptides. Further the antigenic peptide once produced may have to be inaccessible to further processing by being relatively stable or containing a motif that ensures rapid transport into the endoplasmic reticulum where binding to class I protects it from further degradation.

The knowledge of structural features of MHC bound peptides combined with insight in processing and transport mechanisms will allow us to predict antigenic epitopes of pathogens with high accuracy.

Acknowledgements

We thank the Laboratory of Macromolecular Analysis for the use of their facilities and R. Spata for excellent typing. Our work reviewed in this article was supported by the NIH grant 5 R37 AI-07289, 2 P01 AI-10702, 2 P30 CA 13330, and 1 R01 AI-27199.

References

1. Zinkernagel, R.M. and Doherty, P.C. Adv. Immun. 27, 51-177 (1979).
2. Townsend, A.R.M. and Skehel, J.J. J. Exp. Med. 160, 552-563.
3. Townsend, A.R.M., Gotch, F.M. and Davey, J. Cell 42, 457-467 (1985).
4. Townsend, A.R.M. et al. Cell 44, 959-968 (1986).
5. Bjorkman, P.J. et al. Nature 329, 506-512 (1987).
6. Bjorkman, P.J. et al. Nature 329, 512-518 (1987).
7. Garrett, T.P.J., Saper, M.A., Bjorkman, P.J., Strominger, J.L. and Wiley, D.C. Nature 342, 692-696 (1989).
8. Berkower, I., Buckenmeyer, G.K. and Berzofsky, J.A. J. of Immunol. 136, 2498-2503 (1986).
9. Allen, P.M., Matsueda, G.R., Evans, R.J., Dunbar, J.B. Jr., Marshall, G.R. and Unanue, E.R. Nature 327, 713-715 (1987).
10. Rothbard, J.B. and Taylor, W.R. EMBO J. 7, 93-100 (1988).
11. van Bleek, G.M. and Nathenson, S.G. Nature 348, 213-216 (1990).
12. Puddington, L., Bevan, M.J., Rose, J.K., and Lefrancois, L. J. of Virol. 60, 708-717 (1986).
13. Shibata, K., Imarai, M., van Bleek, G.M., Joyce, S. and Nathenson, S.G. Proc. Nat. Acad. Sci. USA, in press.
14. van Bleek, G.M. and Nathenson, S.G. Proc. Natl. Acad. Sci. USA 88, 11032-11036 (1991).
15. Falk, K., Rotzschke, O., Stevanovic, S., Jung, G. and Rammensee, H.-G. Nature 351, 290-296 (1991).
16. Kast, W.M. et al. Cell 59, 603-614 (1989).
17. Schumacher, T.N.M. et al. Cell 62, 563-567 (1990).

5

POSITIVE AND NEGATIVE SELECTION OF T CELLS

Harald von Boehmer

Basel Institute for Immunology
Grenzacherstrasse 487
CH-4005 Basel, Switzerland

INTRODUCTION

The question of self-nonself discrimination by the immune system is still awaiting a complete solution. The problem was recognized a long time ago and Owen's observation[1] (1945) in dizygotic cattle twins, which were chimeric with regard to their blood cells, indicated that self tolerance was acquired rather than inherited. Ideas by Burnet[2] (1957) which were reinforced by Lederberg[3] (1963) had a simple -- possibly too simple -- solution to the problem of self tolerance: it was argued that clones of lymphocytes, each bearing a distinct antigen receptor, went through an early phase in development where contact with antigen was lethal rather than inducing effector function. Thus, tolerance to self was thought to be due to the elimination of self reactive clones early during their development. This concept had two weaknesses in explaining all self-tolerance. First, it was difficult to imagine that all self antigens could reach developing lymphocytes in primary lymphoid organs. Second, considering the enormous diversity of antigen receptors themselves, it was difficult to see how deletion of lymphocytes induced by diverse receptors on other lymphocytes would leave anything substantial behind fit to deal with foreign antigens.[4] Possible solutions to these considerations were first, that clonal deletion was not the only mechanism of tolerance and second, that tolerance induction required a certain threshold amount of antigen which probably was not reached by most idiotypic antigen-receptor sequences. The main question was clearly not whether clonal deletion was the only way to achieve tolerance, but whether clonal deletion existed at all as central tolerance mechanism possibly aided by some additional mechanisms responsible for silencing already matured lymphocytes which had escaped clonal deletion simply because they did not encounter their self antigen early in development.

For decades this question was extremely difficult to address because of the lack of clonotypic receptor markers. This was especially so for T lymphocytes where antigen binding by surface receptors could not be visualized. This left an ambiguity to the interpretation of results obtained with most tolerance models. Tolerance models by Billingham, Brent and Medawar (1953) and Hasek (1953) confirmed that tolerance could be acquired.[5,6] Neonatal tolerance could not be established in all mouse strain combinations, and it was not clear whether the relative ease by which tolerance could be induced in some neonatal mice reflected the immaturity of lymphocytes or simply their relatively small numbers. Conceptually more clearcut were models where hemopoietic stem cells developed in the presence of foreign antigen, in hemopoietic chimeras prepared by injecting stem cells into lethally x-irradiated histoincompatible hosts. In this model, mature T lymphocytes were deliberately removed from the donor cells, and it turned out that this tolerance model worked in all strain combinations. These experiments showed that there was something special about developing lymphocytes with regard to tolerance as they could be adapted to tolerate a large variety of strong histocompatibility antigens.[7] The experiments could, however, not establish that clonal deletion was the mechanism. The readout in all the models depended on activation of lymphocytes *in vivo* or *in vitro* and therefore did not establish whether tolerance was due to silence or absence of lymphocytes bearing specific receptors for the tolerogen. More conclusive studies became only feasible in the mid-1980s, when it was possible to raise antibodies specific for clonotypic receptors on T lymphocytes.[8-10] Even this technological advance was not sufficient to yield conclusive results because the frequency of cells bearing one particular T cell receptor (TCR) was extremely low, such that one clone could not be visualized during lymphocyte development. This was different with cells specific for so-called superantigens which are recognized by particular Vβ protein sequences shared by heterogenous TCRs. Here a relatively large fraction of heterogenous T cells appeared specific for one superantigen.[11] Studies employing superantigens provided some clues to repertoire selection during T cell development[12], but the essential points became clear only when the selection of a single TCR specific for "conventional" antigens, i.e. peptides bound by major histocompatibility complex (MHC) encoded molecules, was studied in T cell receptor transgenic mice.[13] The latter studies clearly established two rules of self-nonself discrimination by the immune system: they showed that tolerance could result from clonal deletion of immature T lymphocytes being confronted with the specific peptide as well as the MHC molecule early in T cell maturation.[14,15] In addition, they showed that T cell maturation required the recognition of the MHC molecule in the absence of the specific peptide early in T cell development.[15-17] As these studies have been published previously, only a brief summary will be provided in the following in order to introduce the experimental system which was used to study further details of intra-thymic selection as well as post-thymic selection of the T cell repertoire.

RESULTS

Negative and positive selection of a transgenic receptor specific for the male-specific peptide presented by class I H-2D^b MHC molecules

The α and β TCR genes from the CD4$^-$8$^+$ cytolytic B6.2.16 clone specific for the male-specific peptide presented by H-2D^b MHC molecules were cloned and cosmids containing flanking sequences which harbored T cell specific regulatory elements (enhancers) were injected into fertilized eggs.[18-20] Transgenic mice which expressed both genes were crossed onto the SCID background in order to produce mice expressing essentially one TCR only.[21] Negative selection (deletion) was studied by comparing male and female $\alpha\beta$ TCR transgenic SCID mice expressing H-2D^b MHC molecules. Positive selection was studied by comparing female $\alpha\beta$ TCR transgenic SCID mice expressing or lacking H-2D^b MHC molecules. The studies involved the analysis of thymocyte subpopulations in the various mice. In an oversimplified way, one can picture the development of thymocytes as going from CD4$^-$8$^-$ thymocytes, which begin to express TCRs, through receptor positive but functionally incompetent CD4$^+$8$^+$ intermediates into CD4$^+$8$^-$ or CD4$^-$8$^+$ TCR positive, functionally competent T cells (Fig. 1).

Figure 2 shows the analysis of the thymus of the experimental mice compared to a normal thymus by staining thymocytes with CD4 and CD8 antibodies. As is evident from Fig. 2, the thymus of male $\alpha\beta$TCR transgenic SCID mice as compared to female littermates is largely devoid of CD4$^+$8$^-$ and CD4$^-$8$^+$ as well as CD4$^+$8$^+$ thymocytes but contains comparable numbers of CD4$^-$8$^-$ thymocytes as female littermates. Because single positive and double positive thymocytes comprise ~90% of all thymocytes, the total number of thymocytes in male animals is only one tenth of that in females. The interpretation of this result is that clonal deletion eliminates even the earliest CD4$^+$8$^+$ immature thymocytes which express the transgenic receptor. These experiments are still consistent with the notion that confrontation with MHC plus peptide leads to an arrest in

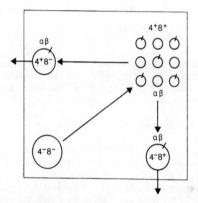

Fig. 1. A scheme of intrathymic development of cells expressing the $\alpha\beta$ T cell receptor: Cycling CD4$^-$8$^-$ lymphoblasts differentiate into noncycling CD4$^+$8$^+$ and finally CD4$^+$8$^-$ and CD4$^-$8$^+$ thymocytes which express the $\alpha\beta$ T cell receptor (symbolized by -).

T cell development rather than physical elimination of immature T cells. Recent in vitro experiments, however, apoptosis could be induced in CD4$^+$8$^+$ thymocytes from the transgenic mice by thymic or peripheral cells expressing D^b MHC molecules plus HY peptide directly show that physical elimination rather than arrest in development accounts for the absence of self reactor T cells in the transgenic mice. The experiments also show that all maturation stages of CD4$^+$8$^+$ cells are susceptible to deletion.[22]

When one compares female $\alpha\beta$TCR transgenic SCID mice which either lack or express H-2D^b MHC molecules, one finds that thymuses from both types of animals contain CD4$^-$8$^-$ as well as CD4$^+$8$^+$ thymocytes, but only those from the latter have CD4$^-$8$^+$ but not CD4$^+$8$^-$ mature thymocytes (Fig. 2). Here the interpretation is that the generation of mature thymocytes requires binding of the $\alpha\beta$TCR to H-2D^b MHC molecules in the absence of the male-specific peptide. Furthermore, binding of the TCR to H-2D^b MHC molecules induces maturation of CD4$^-$8$^+$ thymocytes only, while binding

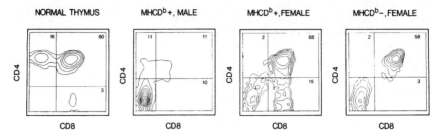

Fig. 2. The subset composition of the thymus from normal mice and three experimental mice by lymphocytes expressing a male specific, D^b-restricted transgenic $\alpha\beta$TCR according to CD4 and CD8 surface markers. Numbers in quadrants represent the percentage of cells in these quadrants.

to class II MHC molecules generates mature CD4$^+$8$^-$ thymocytes only.[23,24] The latter conclusion is well supported by more recent independent experiments employing β2 microglobulin deficient -- and therefore class I MHC molecule deficient -- mice which have CD4$^-$8$^-$, CD4$^+$8$^+$ and CD4$^+$8$^-$ but lack significant numbers of CD4$^-$8$^+$ thymocytes.[25]

Lack of allelic exclusion of the αTCR chain in αβTCR transgenic mice

The picture of thymocyte subsets in $\alpha\beta$TCR transgenic mice which were not crossed with SCID mice was slightly more complicated due to the fact that endogenous a but not b TCR genes could rearrange in these mice. In spite of the fact that the $\alpha\beta$TCR was expressed early on the surface of CD4$^-$8$^-$ thymocytes, endogenous αTCR genes were expressed at later stages at the RNA and protein level such that many cells contained transcripts from both endogenous as well as the transgenic αTCR gene. While in some cases this led to the expression of roughly equimolar amounts of endogenous and transgenic αTCR chains paired with the transgenic βTCR chain, most T cells expressed either high levels of the transgenic or high levels of endogenous αTCR chain on the cell surface presumably due to preferential pairing with the transgenic β chain. In the absence of productive rearrangements of endogenous αTCR gene segments, of course only the transgenic αTCR chain was expressed. These $\alpha\beta$TCR transgenic mice contained significant numbers of CD4$^+$8$^-$ T cells both in the thymus (Fig. 3) as well as the periphery which all expressed high levels of endogenous αTCR chains at the cell surface. In contrast, most CD4$^-$8$^+$ thymocytes expressed high levels of the transgenic αTCR chain and a variable proportion of peripheral CD4$^-$8$^+$ T cells expressed high levels of endogenous αTCR chains. Thus, the early expression of the transgenic αTCR chain did not prevent endogenous αTCR rearrangement. In contrast, the early expression of the β transgene prevented rearrangement of V_β gene segments. Some of the receptors containing endogenous αTCR chains were selectable by thymic class I or class II MHC molecules and therefore, mature T cells bearing these receptors could be detected in the thymus as well as peripheral lymph organs[13] and this has to be taken into consideration in the following experiments which were conducted with $\alpha\beta$TCR transgenic mice which were not backcrossed onto the SCID background.

Positive selection requires expression of MHC molecules on epithelial cells

Experiments in various hemopoietic chimeras constructed by injecting T cell depleted bone marrow cells from female H-2^b mice into lethally x-irradiated recipients of MHC-recombinant and MHC-congenic strains showed that selection of CD4$^-$8$^+\alpha_T\beta_T$

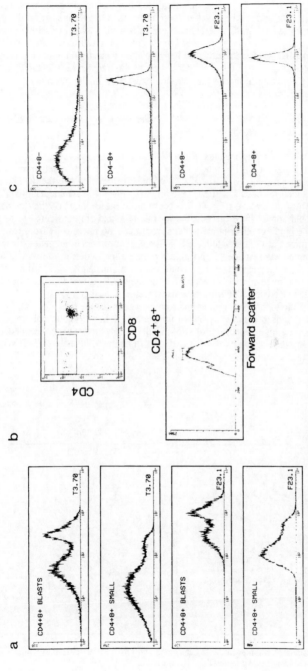

Fig. 3. The subset composition of the thymus from $\alpha\beta$ T cell receptor transgenic mice (not crossed onto the SCID background), by 3-color analysis employing CD4 and CD8 antibodies as well as the F23.1 and T3.70 antibodies specific for the transgenic β and α TCR chains, respectively. Column (a): double positive blast and small cells as defined in column (b). Column (b): CD4/CD8 scattergram and size analysis of double positive cells. Column (c): CD4$^+$8$^-$ and CD4$^-$8$^+$ showing that α chain expression is not subject to allelic exclusion.

49

T cells occurred only in those animals that expressed D[b] MHC molecules on the radio-resistant portion of the thymus.[14] Thus, in spite of the fact that most hemopoietic cells, including dendritic cells and macrophages in these thymuses were derived from the donor stem cells and expressed D[b] MHC molecules, these cells were unable to positively select. These experiments are well in line with experiments employing transgenic animals expressing certain MHC molecules either on hemopoietic or mostly on epithelial cells in the thymus due to differences in the promoter region of the transgenes.[23] Again, positive selection was only observed in mice which showed expression of the relevant MHC molecules on thymic epithelium.

Mutations in the peptide-binding groove of MHC molecules affect antigenicity and negative as well as positive selection.

Two naturally occurring mutants of the D[b] MHC molecule have been isolated, namely in the bm13 and the bm14 mutant strains. The latter contains a single mutation of the inward pointing residue 70 (Glu -> Asp) of the α1 domain while the bm13 contains three changed residues 114 (Leu -> Glu), 116 (Phe -> Tyr) and 119 (Glu -> Asp) at the bottom of the groove (Fig. 4).[26] According to the Bjorkman-Wiley model,[27] these residues would appear to be inaccessible for the TCR and are considered to be involved in peptide binding. In contrast to cells from male B6 mice, cells from male bm13 and bm14 mice fail to stimulate the B6.2.16 clone from which the genes for the male-specific receptor were isolated. This agrees with earlier studies in these mice which showed that cells from female mutant mice could not be stimulated by cells from male littermates to produce male-specific cytolytic T lymphocytes.[28] It was therefore expected that cells with the transgenic αβTCR in mice backcrossed onto the bm13 or bm14 background would show no deletion in animals homozygous for the bm13 or bm14 mutation. This was indeed the case as shown for the bm13 homozygous male mice in Fig. 5 which contain a high proportion of CD4[+]8[+] thymocytes expressing intermediate levels of the transgenic receptor on the cell surface. In contrast, in male D[b]/bm13 heterozygous mice this population is entirely deleted.[26]

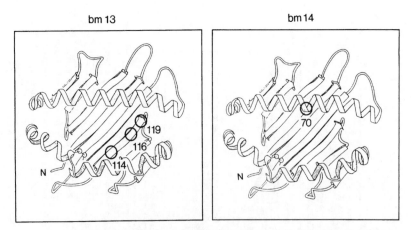

bm 13   bm 14

Fig. 4. Position of residues in the mutant bm13 and bm14 MHC molecules which differ from the wildtype D[b] MHC molecule.

As positive selection proceeds in the absence of the specific peptide, one may or may not have anticipated that MHC mutations restricted to the peptide-binding groove would not affect positive selection. The fact that MHC molecules seldom exist on the surface of cells without a bound peptide[29] creates difficulties with the view that positive selection would simply be achieved by the binding of TCR to upwards pointing residues of MHC molecules without the positive or negative interference of the MHC-bound peptides. It may therefore not be surprising that indeed, as a rule, mutations which affect antigenicity also affect positive selection.[26,30,31] This is shown in Fig. 6 for the bm14

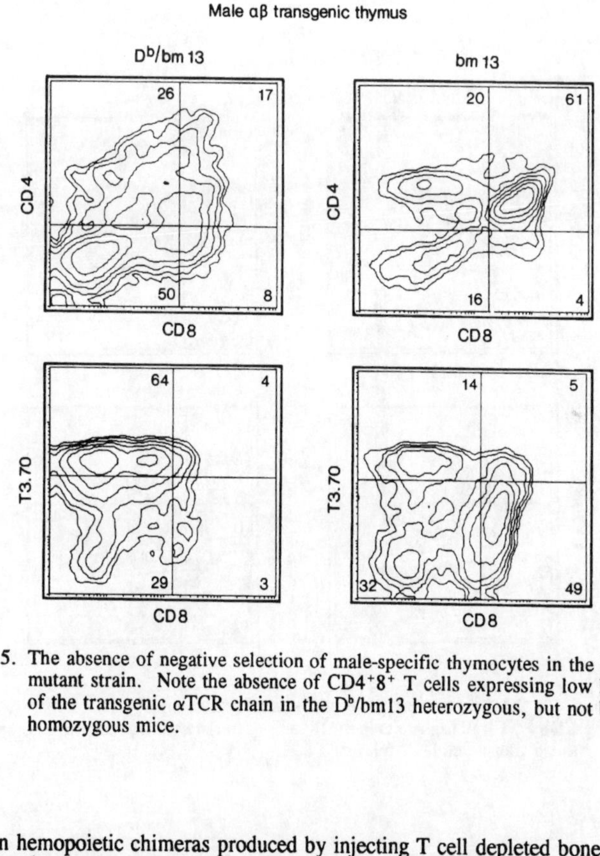

Male αβ transgenic thymus

Fig. 5. The absence of negative selection of male-specific thymocytes in the bm13 mutant strain. Note the absence of CD4$^+$8$^+$ T cells expressing low levels of the transgenic αTCR chain in the D^b/bm13 heterozygous, but not bm13 homozygous mice.

mutation: in hemopoietic chimeras produced by injecting T cell depleted bone marrow cells from H-2^b αβTCR transgenic female mice into lethally x irradiated H-2D^b(B6) homozygous or bm14 homozygous recipient mice, it is evident that positive selection of CD4$^-$8$^+$, T3.70 cells occurs in the former but not the latter recipients. Unfortunately, these experiments do not tell us whether the role of the peptide in positive selection is facilitating or interfering with positive selection by MHC molecules; they simply tell us that peptides are not ignored during positive selection. Alternatively, they could also mean that mutations in the peptide binding groove do affect the overall structure of the MHC molecule.

Positive selection: expansion of mature thymocytes or maturation of immature thymocytes?

We have addressed the question whether positive selection is reflected in expansion of mature T cells rather than maturation of immature resting cells by labelling experiments employing bromodeoxyuridine (BrdU) which incorporates into newly synthesized DNA and can be visualized by a monoclonal antibody specific for BrdU.[32] Continuous labelling experiments (BrdU injected i.p. every day early in the morning and late in the evening) show that in female H-2^b αβTCR transgenic mice, most CD4$^+$8$^+$ but

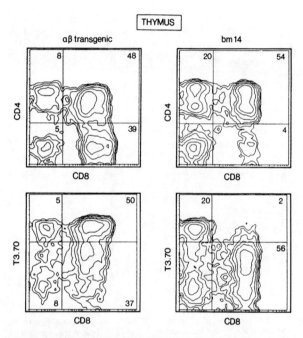

Fig. 6. Lack of positive selection in bm14 homozygous mice. Note the absence of CD8+, T3.70 high cells in the thymus of bm14 homozygous mice compared to αβ transgenic H-2^b mice.

none of the CD4$^-$8$^+$ thymocytes are labelled after three to four days of continuous labelling. This indicates that CD4$^-$8$^+$ thymocytes do not divide in these mice (Fig. 7). Whereas the label accumulates linearly in the CD4$^+$8$^+$ compartment during the first five days, it accumulates in nonlinear fashion in the CD4$^-$8$^+$ compartment from day 4 onwards. If one compares the absolute numbers of labelled CD4$^+$8$^+$ and CD4$^-$8$^+$ cells in αβ transgenic and normal B6 mice, it is clear that in the former the transition of CD4$^+$8$^+$ precursors into CD4$^-$8$^+$ progeny is much more efficient than in the latter (Fig. 7).[33]

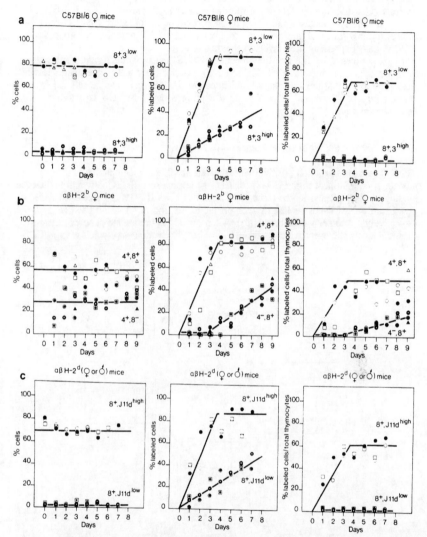

Fig. 7.   The increase in percentage of labelled CD4⁺8⁺ and CD4⁻8⁺ cells during continuous labelling of **(a)** female normal B6 mice, **(b)** $\alpha\beta$TCR transgenic H-2^b mice, and **(c)** $\alpha\beta$TCR transgenic H-2^d mice.   On the left hand side the proportion of cells in each subset throughout the duration of the labelling experiment is shown.   In the middle is the proportion of labelled cells in each subset, and on the right is the proportion of labelled cells per total thymocytes.

An independent experimental protocol leads to the same conclusion: if one reconstitutes x-irradiated H-2^b mice with varying mixtures of T cell depleted bone marrow cells from normal female B6 mice as well as H-2^b female $\alpha\beta$TCR transgenic mice, one observes that at a ratio of 1:20 the proportion of immature CD4$^+$8$^+$ cells expressing low levels of the transgenic receptor is around 5% whereas the proportion of CD4$^-$8$^+$ cells expressing high levels of the transgenic receptor is more than 70% (Fig. 8). These experiments indicate that the immature CD4$^+$8$^+$ cells expressing the transgenic receptor mature preferentially into CD4$^-$8$^+$ cells. Collectively these experiments leave very little doubt that the vast majority of immature CD4$^+$8$^+$ cells can be rescued from programmed cell death and induced to maturation provided these cells express a selectable receptor.

Commitment to the CD4 or CD8 lineage: instruction versus selection

The experiments on positive selection were in principle compatible with two models of commitment to the CD4$^+$8$^-$ or CD4$^-$8$^+$ lineage of cells. One would be that the binding of the TCR and the co-receptor to class I or class II MHC molecules in the thymus instructs CD4$^+$8$^+$ cells to switch off the expression of CD4 and CD8 genes, respectively. Another would be that the switch off of CD4 and CD8 gene expression is an initial stochastic event which is then followed by a selection step which rescues CD8$^+$

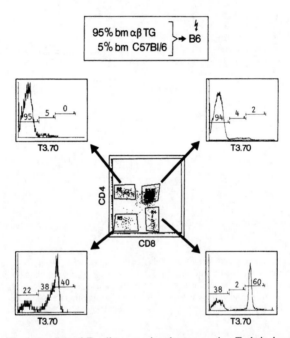

Fig. 8. The proportion of T cells expressing the transgenic αT chain in various thymocyte subsets in chimeras which were constructed by injecting bone marrow cells from normal B6 and $\alpha\beta$TCR transgenic female H-2^b mice at a ratio of 1 to 20 into lethally x-irradiated B6 female recipient mice.

cells with class I and CD4$^+$ cells with class II MHC restricted receptors while CD8$^+$ cells with class II and CD4$^+$ cells with class I MHC restricted receptors would die. In its simplest form the selective model predicts that a CD8 gene, which because of regulation by a T cell specific promoter can be expressed in the CD4$^+$ as well as the CD8$^+$ lineage of cells should rescue CD4$^+$ cells with a class I MHC restricted receptor from cell death. We have performed such an experiment in double transgenic mice expressing a transgenic class I specific $\alpha\beta$TCR as well as CD8 gene(s) and did not find any evidence supporting the selective model.[34,35] During the course of these studies we discovered, however, that female mice which expressed the HY peptide plus D^b MHC specific receptor and D^b MHC molecules had a much higher proportion of CD4$^+$8$^+$ T cells expressing high levels of the TCR than D^b-negative $\alpha\beta$TCR transgenic mice. By a careful comparison of these cells in D^b-positive and D^b-negative strains we found out that CD4$^+$8$^+$ TCR high cells are the result of positive selection. These cells can be characterized by a variety of markers which all suggest that they are intermediates between CD4$^+$8$^+$ TCR^{low} and CD4$^-$8$^+$ single positive cells. In fact, when these cells were cultured in vitro after intensive purification by repeated cell sorting, we could show that they developed into functional CD4$^-$8$^+$ T cells in the absence of positively selecting ligands on thymic epithelial cells. Thus, these experiments favor the instructional model of lineage commitment in that they show that CD4$^+$8$^+$ thymocytes are already determined to develop into one or the other subset according to the specificity of their receptor.

Post-thymic expansion of mature T cells.

Once mature T cells are formed they can leave the thymus and undergo considerable expansion in peripheral lymph organs. The expansion potential of T cells has been shown in a variety of experiments, probably most directly by transferring small numbers of peripheral T cells into T cell-deficient nu/nu mice lacking a thymus, where both CD4$^+$8$^-$ and CD4$^-$8$^+$ expand considerably.[36] It is not clear whether this peripheral expansion randomly amplifies patterns of TCR specificities as they were selected in the thymus or whether it is not random and modifies patterns which were initially selected in the thymus. To study these questions, mature T cells from female $\alpha\beta$TCR transgenic mice were transferred into female and male nu/nu recipients. The remarkable

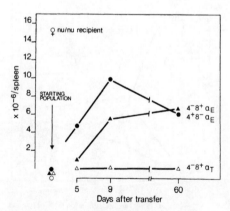

Fig. 9. The total number of T cells in spleen and lymph nodes from female nu/nu mice injected with T cells from female $\alpha\beta$TCR transgenic H-2^b mice. α_E means endogenous αTCR chains, α_T transgenic αTCR chains.

observation in these experiments was that the entire increase in T cells in the T cell deficient recipients could be attributed to CD4$^+$8$^-$ and CD4$^-$8$^+$ T cells expressing endogenous αTCR chains even though approximately half of the inoculated CD4$^-$8$^+$ T cells expressed the male-specific transgenic αβTCR. In fact, these male-specific cells did not expand at all and the total number of cells with this phenotype was the same at day 5 and day 60 after transfer (Fig. 9). This indicates that post-thymic repertoire selection is totally different from thymic selection with regard to both specificity and mechanism: the former selects specificities of non-transgenic TCRs and involves expansion of cells, the latter selects predominantly the specificity of the transgenic TCR and involves maturation rather than expansion. The experiments also indicate that naive, mature T cells, once they leave the thymus, have a relatively long lifespan in peripheral lymphoid tissue (up to two months). This has, in the meantime, been confirmed in αβTCR transgenic female mice by conducting labelling experiments with DNA precursors.

Post-thymic tolerance.

We have begun to address the induction of unresponsiveness in mature T cells by transferring T cells from female αβTCR transgenic B6 mice into male nu/nu B6 mice and by following the fate of male-specific T cells in these animals. As shown in Fig. 10, these cells proliferate initially vigorously such that by day 5 after transfer the number of peripheral T cells is four times higher than that observed in female recipients injected with the same numbers of cells. After this, the number of male-specific CD4$^-$8$^+$ T cells declines steadily but significant numbers can still be detected even eight weeks after transfer. When these cells are analyzed more carefully, one finds that their size is equal to that of cells with the transgenic receptor in female mice but that the levels of CD8 co-receptors as well as that of the transgenic receptor is lower in the male compared to the female recipients. In fact, the cells from the male recipients no longer proliferate in response to antigenic stimulation by male cells in vitro, even in the presence of IL2, while proliferation of such male-specific cells is easily detected with cells from female recipients (Table 1). Thus, as post-thymic "positive selection" differs from thymic "positive selection", so does thymic "negative selection" differ from post-thymic "negative selection". Post-thymic "negative selection" involves initial expansion and

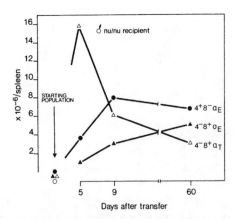

Fig. 10. The total number of T cells in spleen and lymph nodes from male nu/nu mice injected with T cells from female αβTCR transgenic H-2^b mice.

Table 1. Proliferative response of T cells from nu/nu mice injected with T cells from transgenic mice.

cells	nu/nu ♂ nu/nu ♀ + IL2		nu/nu ♂	nu/nu ♀
♀ αβ CD8	150'000	4'800	37'400	600
♀ αβ CD8 from nu/nu♂ d5	74'500	35'200	17'500	5'000
♀ αβ CD8 from nu/nu♂ d9	42'300	46'000	1'000	1'700
♀ αβ CD8 from nu/nu♂ d20	7'400	5'800	600	700

5×10^4 responder cells (B cell and CD4 cell depleted lymph node cells) from various mice were cultured with 5×10^5 x-irradiated (3000R) spleen cells from male or female nude mice in the presence or absence of exogenous interleukin 2. Values represent the mean cpm of ^{3}H-thymidine of triplicate cultures. Standard errors = or < 10%.

subsequent elimination of cells and/or down-regulation of antigen receptors as well as co-receptors, thymic "negative selection" involves deletion of cells at an immature stage without prior expansion.

DISCUSSION

Monoclonal antibodies specific for clonotypic T cell receptors (TCR) and TCR transgenic mice have made it possible to study thymic[13-23] and post-thymic T cell repertoire selection in previously unknown detail. The conclusions are that T cell maturation depends on "assertion" of self, namely the binding of the $\alpha\beta$TCR to thymic MHC molecules in the absence of the specific peptide recognized by the receptor when expressed on mature T cells. This event is called "positive selection", rescues CD4$^+$8$^+$ thymocytes from programmed cell death[37,38] and induces maturation into mature single positive thymocytes in the absence of cell division. An important feature of this selective step is that binding of the $\alpha\beta$TCR to class I or class II MHC molecules determines that the mature T cells is of the CD4$^-$8$^+$ and CD4$^+$8$^-$ phenotype, respectively.[16,17,23,24]

In the presence of the specific peptide, one observes the opposite: CD4$^+$8$^+$ thymocytes are not even allowed to persist throughout their programmed lifespan but they are deleted as soon as they start to express the CD4/CD8 co-receptors.[14,15] This event is called "negative selection" and prevents the entry of autoaggressive T cells into the peripheral lymphocyte pool.

Once the cells have passed through positive selection and avoided negative selection, they can expand considerably in peripheral lymph organs.[36,39] This expansion depends again on the specificity of the $\alpha\beta$TCR but in a different way than when compared with intra-thymic positive selection: here the receptor must bind to both the peptide as well as the presenting MHC molecule. The consequence of "positive selection" in peripheral lymph organs is also different from that in the thymus: in the former it is expansion and maturation into effector cells, while in the latter the consequence is differentiation into mature lymphocytes.

As predicted by Burnet[2] (1957) and also by Lederberg[3] (1963), "negative selection" of mature T cells in peripheral or secondary lymph organs differs from that in primary lymph organs: in the thymus cells can be deleted without prior stimulation to expand while in secondary lymph organs, cells expand considerably but then either disappear and/or down-regulate antigen receptors and co-receptors such that the cells become refractory to antigenic stimulation even in the presence of exogenous interleukin 2. This mode of peripheral tolerance which we observe in male nu/nu mice injected with male-specific cells from $\alpha\beta$TCR transgenic mice differs from that seen in transgenic animals where certain antigen are expressed exclusively by "nonprofessional antigen" presenting cells; in these experiments the antigens induce result a state of anergy in antigen-specific T cells which can be overcome by the addition of exogenous interleukin 2.[40] Thus, in addition to one central tolerance mechanism, there appear to be at least two and possibly more peripheral tolerance mechanisms, hopefully including some which are mediated by so-called suppressor cells.

While the biological sense of negative selection of potentially autoaggressive cells is obvious, the biological meaning of positive selection has confused immunologists to some extent:[41] the high frequency of alloreactive T cells, many of which probably recognize some peptides presented by foreign MHC antigens, in a repertoire selected by self MHC antigens appears to make positive selection superfluous. It has been demonstrated in the past by experiments employing hemopoietic chimeras that this is not true: experiments in x-irradiated, MHC homozygous animals reconstituted with cells from donors expressing either completely allogeneic or semi-allogeneic MHC molecules have shown that positive selection has a considerable impact on the "immunological fitness": immune responses in allogeneic chimeras (as long as one does not study responses to a multitude of allogeneic major or minor histocompatibility antigens)[42] are very much reduced even after in vivo priming (which, as we have shown in the transfer experiments reported here, can lead to enormous expansion of peripheral T cell); likewise, responses "restricted" by MHC molecules not encountered in the thymus are very much lower, or even absent, in semi-allogeneic chimeras.[43] Thus, after negative selection, responses restricted by MHC molecules not encountered in the thymus are, as a rule, orders of magnitude lower than those restricted by MHC molecules which were encountered in the thymus.

But does positive selection increase the frequency of self MHC restricted cells compared to the unselected repertoire? This question is difficult to address experimentally but the following consideration may indicate that it is so. From the data on positive selection of several transgenic $\alpha\beta$TCRs which have been studied so far, it is apparent that different MHC molecules rarely select the same $\alpha\beta$TCR and thus, in a first approximation, one can look at the unselected repertoire of immature T cells as a collection of fifty or so repertoires which can be selected by fifty different MHC molecules. Considering this assumption and the data in allogeneic or semi-allogeneic chimeras, one comes to the conclusion that in the unselected repertoire, the frequency of self MHC restricted T cells will always be lower than in the selected repertoire. Thus, positive selection serves to increase the "immunological fitness" of the mature T cell pool in spite of some confusion raised by earlier experimental data. In summary, one may conclude that the experimental demonstration of allo-restricted T cells has not been a good argument against the beneficial effect of positive selection, but that the low frequency of T cells restricted by MHC antigens not present on the thymus in allogeneic or semi-allogeneic chimeras has documented the usefulness of positive selection.

REFERENCES

1. Owen, R.D. *Science* 102:400 (1945).
2. Burnet, F.M. *Aust. J. Sci.* 20:67 (1957).
3. Lederberg, J. *Science* 129:1649-1653 (1959).
4. Jerne, N.K. *Ann. Immunol.* 125:373 (1974).
5. Billingham, R.E., Brent, L., Medawar, P.B. *Nature* 172:603 (1953).
6. Hasek, M. *Cechoslovenska Biol.* 2:265 (1953).
7. von Boehmer, H., Sprent, J., Nabholz, M. *J. Exp. Med.* 141:322 (1975).
8. Allison, J.P., McIntyre, B.W., Bloch, D. *J. Immunol.* 129:2293 (1982).
9. Meuer, S.C., Fitzgerald, K.A., Hussey, R.E., Hodgdon, J.C., Schlossman, S.F., Reinherz, E.L. *J. Exp. Med.* 157:705 (1983).
10. Haskins, K., Kubo, R., White, J., Pigeon, M., Kappler, J.W., Marrack, P.C. *J. Exp. Med.* 157:1149 (1983).
11. Marrack, P., Kappler, J. *Science* 248:705 (1990).
12. Kappler, J.W., Roehm, N., Marrack, P.C. *Cell* 49:273-280 (1987).
13. von Boehmer, H. *Ann. Rev. Immunol.* 8:531 (1990).
14. Kisielow, P., Teh, H.S., Blüthmann, H., von Boehmer, H. *Nature* 335:730 (1988).
15. Sha, W.C., Nelson, C.A., Newberry, R.D., Kranz, D.M., Russell, J.H., Loh, D.Y. *Nature* 335:271 (1988).
16. Teh, H.S., Kisielow, P., Scott, B., Kishi, H., Uematsu, Y., Blüthmann, H., von Boehmer, H. *Nature* 335:229 (1988).
17. Kisielow, P., Blüthmann, H., Staerz, U.D., Steinmetz, M., von Boehmer, H. *Nature* 335:229 (1988).
18. Uematsu, Y., Ryser, S., Dembic, Z., Borgulya, P., Krimpenfort, P., Berns, A., von Boehmer, H., Steinmetz, M. *Cell* 52:831 (1988).
19. Krimpenfort, P., de Jong, R., Uematsu, Y., Dembic, Z., Ryser, S., von Boehmer, H., Steinmetz, M., Berns, A. *EMBO J.* 7:745 (1988).
20. Blüthmann, H., Kisielow, P., Uematsu, Y., Malissen, M., Krimpenfort, P., Berns, A., von Boehmer, H., Steinmetz, M. *Nature* 334:156 (1988).
21. Scott, B., Blüthmann, H., Teh, H.S., von Boehmer, H. *Nature* 338:591 (1989).
22. Swat, W., Ignatowicz, L., von Boehmer H. and Kisielow, P. *Nature* 351:150 (1991).
23. Berg, L.J., Pullem, A.M., Fazekas de St Groth, B., Mathis, D., Benoist, C., Davis, M.M. *Cell* 58:1035-1046 (1989).
24. Kaye, J., Hsu, M.L., Sauron, M.E., Jameson, J.C., Gascoigne, R.J., Hedrick, S.M. *Nature* 341:746-749 (1989).
25. Zijlstra, M., Bix, M., Simister, N.E., Loring, J.M., Raulet, D.H., Jaenisch, R. *Nature* 344:742 (1990).
26. Jacobs, H., von Boehmer, H., Melof, C.J.M., Berns, A. *Eur. J. Immunol.* (in press, 1992).
27. Bjorkman, P., Saper, M., Samraoui, B., Bennet, W., Strominger, J., Wiley, D. *Nature* 329:512-518 (1987).
28. de Waal, L.P., Melvold, R.M., Melof, C.J.M. *J. Exp. Med.* 158:1537 (1983).
29. Ljunggren, H., Stam, N.J., Oehlen, C., Neefjes, J.J., Höglund, P., Heemels, M., Bastin, J., Schumacher, T.N.M., Townsend, A., Kärre, K., Ploegh, H.L. *Nature* 346:476 (1990).
30. Nikolic-Zugic, J., Bevan, M.J. *Nature* 344:65-67 (1990).
31. Sha, W.C., Nelson, C.A., Newberry, R.D., Pullen, J.K., Pease, L.R. Russell, J.H., Loh, D.Y. *Proc. Natl. Acad. Sci. USA* 87:6186 (1990).
32. Dolbeere, F., Gretner, H., Pallovicini, M.G., Gray, J.V. *Proc. Natl. Acad. Sci. USA* 80:5573 (1983).
33. Huesmann, M., Scott, B., Kisielow, P. and von Boehmer, H. *Cell* 66:533 (1991).
34. Robey, E.A., Fowlkes, B.J., Gordon, J.W., Kioussis, D., von Boehmer, H., Ramsdell, F. and Axel, R. *Cell* 64:99 (1991).

35. Borgulya, P., Kishi, H., Müller, V., Kirberg, J. and von Boehmer, H. *EMBO J.* 10:913 (1991).
36. Rocha, B., Dautigny, N., Pereira, P. *Eur. J. Immunol.* 19:905-911 (1989).
37. von Boehmer, H. *Immunol. Today* 7:333 (1986).
38. von Boehmer, H. *Ann. Rev. Immunol.* 6:309-326 (1988).
39. Miller, R., Stutman, O. *J. Immunol.* 133:2925-2932 (1984).
40. Lo, D., Burkly, L., Flavell, R., Palmiter, R., Brinster, R.J. *J. Exp. Med.* 169:779 (1989).
41. Matzinger, P. *Nature* 292:497 (1981).
42. Matzinger, P., Mirkwood, G. *J. Exp. Med.* 184:84 (1978).
43. von Boehmer, H., Teh, H.S., Bennink, J.R., Haas, W. *In*: "Recognition and Regulation in Cell Mediated Immunity," J.D. Watson and J. Mabrook, ed., Dekker, New York and Basel (1985), p. 89.

6

PERIPHERAL TOLERANCE

David C. Parker

Department of Molecular Genetics and Microbiology
University of Massachusetts Medical School
Worcester, MA 01655

THE PROBLEM

The problem of self-tolerance is far from a satisfactory
solution. The dominant idea for decades has been the deletion
of self-reactive lymphocytes when they first express their
unique antigen receptors[1]. Clonal deletion has recently been
dramatically verified for both T cells in the thymus[2,3] and B
cells in the bone marrow[4], but we know that clonal deletion is
incomplete. Deletion of B cells depends on crosslinking of B
cell surface immunoglobulin, and does not occur with soluble
self proteins[5]. Deletion of T cells in the thymus requires
presentation of self peptides at adequate levels in the thymus,
a mechanism which may be limited to several hundred self
peptides by the number of available MHC molecules on deleting
thymic antigen presenting cells (APCs)[6]. The existence of
self-reactive lymphocytes in the periphery is demonstrated by
the immune responses and experimental autoimmune diseases which
can be induced by immunizing with self antigens in strong
adjuvants. The mechanisms which limit the immune responses of
potentially autoreactive lymphocytes once they leave the bone
marrow and the thymus are undoubtedly complex, and probably
involve every regulated step of the immune response.

This overview will focus on the decision of a potentially
autoreactive T cell when it first encounters antigen after it
leaves the thymus: how does it distinguish self from not-self
(Figure 1)?

APC FUNCTION IS TIGHTLY REGULATED

Since the T cell receptor is already selected, the T cell
cannot rely on its clonal specificity to distinguish self from
not-self. It must rely on other clues in its local environment,
particularly the APC. As Janeway has pointed out[7], T cells,
particularly virgin T cells, are difficult to activate, and
immunologists are used to relying on adjuvants to stimulate
vigorous immune responses. Adjuvants mimic the effects of local

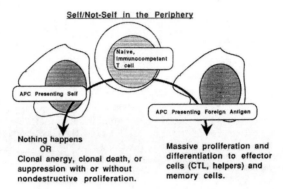

Self/Not-Self in the Periphery

Naive, immunocompetent T cell

APC Presenting Self

APC Presenting Foreign Antigen

Nothing happens
OR
Clonal anergy, clonal death, or suppression with or without nondestructive proliferation.

Massive proliferation and differentiation to effector cells (CTL, helpers) and memory cells.

Figure 1. The decision of a virgin, immunocompetent T cell which exits the thymus with a self-reactive antigen receptor. How does it distinguish self from not-self?

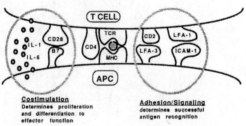

APC Function Is Regulated

1. Antigen presentation:
   • antigen uptake
   • MHC expression
2. Accessory molecules
3. Costimulation

T CELL

IL-1
IL-6
CD28
B7
TCR
CD4
MHC
CD2
LFA-1
LFA-3
ICAM-1

APC

Costimulation
Determines proliferation and differentiation to effector function

Adhesion/Signaling
determines successful antigen recognition

Figure 2. APC functions.

infection, by provoking aggregation of antigen, maintaining high local antigen concentrations, and incorporating bacteria or bacterial products that stimulate the primitive receptors of innate immunity to upregulate the antigen presenting function of APCs.

Antigen presentation is a complex process, each step of which is regulated (Figure 2). The first step is antigen uptake, which depends on whether antigen is particulate or soluble or infectious. Antigen uptake can be dramatically enhanced by complexing of antigen with antibody or complement[8]. The next step is antigen processing: the creation of a peptide/MHC complex that serves as a ligand for the T cell antigen receptor. Levels of expression of MHC molecules are constitutively low in most cell types, and are upregulated by inflammatory cytokines and lymphokines. Since MHC molecules may typically be loaded with peptide only once in the lifetime of each MHC molecule, it is likely that MHC synthetic rate determines the efficiency of antigen processing (Lanzavecchia, this volume). Successful recognition of the peptide/MHC complex on the surface of the APC requires the participation of additional surface molecules involved in adhesion and signalling. In addition to the coreceptors, CD4 and CD8, adhesion/signalling ligand pairs which determine successful antigen recognition include CD2/LFA-3 and LFA-1/ICAM-1[9] (Figure 2). These are also upregulated by lymphokines and cytokines. Other regulated adhesion systems determine lymphocyte trafficking and access to various tissue compartments.

However, antigen recognition is not sufficient to result in T lymphocyte proliferation and differentiation to helper or cytotoxic function. The APC needs to provide additional signals which collectively are termed costimulation. These include soluble cytokines acting locally like IL-1 and IL-6 as well as membrane bound molecules which deliver signals to membrane bound receptors on the T cell. The best example of the latter is the interaction between B7, an activation antigen of B cells and macrophages, and CD28, a molecule which enhances IL-2 secretion by T cells using cytoplasmic signalling pathways which are distinct from those used by the the T cell antigen receptor[10,11]. Costimulatory signals differ among T lymphocytes in different states of differentiation. Naive T cells have different requirements from memory T cells[12]. Differences also have been found between functional CD4 subsets[13,14] and Th1 and Th2 continuous T cell lines[14,15].

Costimulatory signals must also be induced. In the response of naive T cells to alloantigens, only specialized dendritic cells constitutively express high levels of costimulatory signals necessary to induce proliferation in the mixed lymphocyte reaction[16]. Dendritic cells may also be necessary to prime naive T cells to protein antigens. The B7 costimulatory molecule is expressed on resting but not activated B cells, and is induced on macrophages by interferon-γ[10]. Activated macrophages also release IL-1. IL-6, produced by many cell types, is a major mediator of inflammation.

CONSEQUENCES OF ANTIGEN RECOGNITION WITHOUT COSTIMULATION

What happens to the T cell when it encounters self antigens on healthy tissues? The most likely outcome is no effect, because of lack of effective antigen processing or recognition. But there is much recent evidence in support of an old theory[17,18] that effective antigen recognition (signal 1) without essential costimulatory signals (signal 1) results in T cell unresponsiveness (anergy) or death. Murine Th1 cell lines that recognize antigen without the appropriate cell-associated costimulatory signal (probably B7) enter a stable anergic state in which they cannot proliferate in response to antigen with costimulation[19]. Induction of anergy in Th1 cell lines requires new RNA and protein synthesis and is sensitive to inhibition by cyclosporin A. Anergy in this system affects IL-2 gene transcription and little else, with no obvious changes in the rest of the cellular machinery for antigen recognition and signal transduction. Anergic cells fail to make IL-2, but do make reduced amounts of other lymphokines. Anergy can be induced by increasing intracellular calcium with calcium ionophores, or by blocking the autocrine IL-2 loop with anti-IL-2 and anti-IL-2 receptor even in the presence of the costimulatory signal[20]. It can be reversed by several rounds of cell division in response to exogenous IL-2.

Anergic cells have also been produced in vivo by transfer of mature, reactive T cells into an allogeneic environment[21-23], and characterized in transgenic animals bearing alloantigens expressed extrathymically under the control of tissue-specific promoters[24] and in animals transgenic for a self-reactive T cell antigen receptor[25]. Some of these in vivo anergic cells resemble the anergic Th1 lines in being incapable of producing IL-2 while remaining IL-2 responsive, but others have quite distinct phenotypes[25], and it will be some time before the various forms of T cell anergy receive precise definitions[26]. The murine T cell lines provide a system in which the induction of anergy and the anergic state can be defined biochemically. But the crucial encounter in self/not-self discrimination would seem to be that of the naive T cell with antigen. The low frequency of T cells specific for any one antigen creates difficulties for the study of the requirements for induction of anergy or unresponsiveness in naive T cells in vitro. These can be circumvented by studying responses to alloantigens or superantigens[27,28] or anti-CD3, or the use of T cells from T cell antigen receptor transgenics. The many models of transgenic animals expressing foreign antigens from transgenes in different tissues and cell types will also provide very useful systems for the study of self-tolerance in the periphery[29].

ACQUIRED TOLERANCE AND SUPPRESSION

Insight into mechanisms of self-tolerance in the periphery can be gained by studying acquired immunological tolerance, in which exposure to antigen under certain conditions results in specific immunological unresponsiveness rather than priming for a secondary response. The conditions for tolerance induction (soluble, deaggregated antigen without adjuvants, or APC-depleted tissue) would allow for antigen recognition without induction of costimulatory signals, and so are likely to induce tolerance by clonal inactivation as a result of signal 1 without signal 2. An alternative explanation for acquired tolerance is the induction of suppressor T cells, which specifically recognize antigen or antigen receptors on antigen responsive cells and block a specific immune response[30]. In the context of self-tolerance, suppressor cells would constitute a dominant self-recognition system which would suppress the responses of potentially destructive autoreactive cells[31,32].

Clonal anergy and suppression are not mutually exclusive concepts. It has been proposed that anergic cells persist to act as self-reactive suppressor cells to block the proliferation and differentiation of other, potentially destructive autoreactive clones[33,34]. Suppression by self-reactive, anergic T cells may be particularly important to block or break a positive feedback loop in which a proliferating, potentially destructive clone induces its own costimulatory functions. Therefore, one might expect that suppression would act on the APC to block costimulation when costimulation is being upregulated solely by T cell derived lymphokines, and not other mediators of inflammation associated with infection or strong adjuvants.

B CELLS AS TOLERIZING APC

In vitro experiments in my laboratory showed clearly that small B cells are excellent APC to induce help from cloned T cell lines[35], but others have shown that small B cells are defective APC for primary T dependent responses in vitro and in vivo[16,36,37], and can induce tolerance to alloantigens[38]. We reasoned that they might be defective APC for primary responses because they lack costimulatory accessory molecules[18], and might produce unresponsiveness in the T cell compartment by delivering signal 1 without signal 2, as in the anergy model in Th1 cell lines[19]. We proposed that B cells act as antigen-specific, tolerogenic APC in acquired tolerance to protein antigens, when antigens are given in low doses in soluble form. B cells are extraordinarily efficient antigen presenting cells for antigens which they can bind with their antigen receptors. Antigen-specific B cells can present antigens at concentrations ten thousand-fold lower than other APC in spleen[8,39]. Under conditions of tolerance induction with repeated low concentrations of soluble antigens[40], the rare, antigen-specific B cells may be the only APC which can collect enough antigen to affect the T cell in any way.

We devised two experimental systems to test the hypothesis that resting B cells act as antigen-specific APC to induce tolerance to protein antigens in naive T cells[41,42]. In the first system, antigen (Fab fragments of rabbit IgG) is targeted directly to small B lymphocytes in vivo by intravenous injection of Fab fragments of rabbit anti-IgD antibody. We use monovalent Fab fragments to avoid activating the B cells by crosslinking membrane IgD, which results in a vigorous polyclonal response in vivo[43]. In the second system, B cells from transgenic mice expressing the membrane form but not the secreted form of human μ chain are transferred to normal littermates, and then the ability to mount an antibody response to human μ chains (Fcμ fragment) is measured. In each case, treated animals are rendered specifically tolerant as measured by serum antibody titers following challenge with antigen in adjuvant. In the anti-IgD system, adoptive transfer shows that tolerance is induced in the T cell compartment, and proliferative responses of primed T cells from treated animals are also reduced. These experiments support a role for B cells as antigen-specific antigen presenting cells in acquired tolerance to protein antigens, and suggest a role for B cells in self-tolerance to soluble proteins.

Since B cells specific for soluble self antigens are not deleted, but persists as responsive[44] or anergic, antigen-binding cells[5], they are constitutively presenting self antigens ten thousand-fold more efficiently than other, antigen-nonspecific APC, including the deleting dendritic cells and macrophages in the thymus. It follows that immunocompetent T cells leaving the thymus will encounter certain soluble self antigens presented only on rare, self antigen-specific B cells. If these B cells induce tolerance in naive T cells, as we would predict from the experiments described above, such T cells specific for low concentration self antigens will be inactivated. Self-reactive B cells will dramatically lower the concentration threshold for self-tolerance to soluble self proteins. If the B cells were to induce proliferation of naive T cells and differentiation to helper function, the result would be autoantibody production, unless the B cells were anergic. If they had no effect on the naive T cells, such a T cell might cause trouble later, were it to encounter a foreign antigen on a competent APC and differentiate into a clone of helper effector T cells. These T cells would be likely to cause autoantibody formation when they encounter autoreactive B cells presenting the cross-reactive self peptide once more.

Trouble could result also when the self-reactive B cell participates in an antibody response to a cross-reactive foreign antigen. As long as help for this B clone clone depends on a T cell epitope in the foreign antigen, autoantibody production will cease when the antigen is cleared. But the activated B cell will also present self peptide, this time with its costimulatory signals induced. If helper T cells specific for the self peptide had not been inactivated by a previous encounter with the B cell in its resting, tolerizing mode, a self-perpetuating autoantibody response could follow. Activated B cells can drive helper T cell proliferation in vitro and in vivo[45], although it is not known whether they can prime naive T cells. Recently, Lin and Janeway[46] have shown directly that induction of autoreactive B cells that cross-react with a foreign antigen allows priming or expansion of non-cross-reactive, autoreactive T cells by this mechanism. It seems likely that similar mechanisms involving cross-reactions with environmental antigens or pathogens may account for the breakdown of self-tolerance in the induction of naturally occurring autoimmune disease.

REFERENCES

1. Lederberg, J. 1959. Genes and antibodies. *Science* 129:1649.
2. Kappler, J.W., N. Roehm, and P. Marrack. 1987. T cell tolerance by clonal elimination in the thymus. *Cell* 49: 273.

3. Kisielow, P., H. Bluthmann, U.D. Staerz, M. Steinmetz, and H. von Boehmer. 1988. Tolerance in T cell receptor transgenic mice involves deletion of nonmature CD4+CD8+ thymocytes. *Nature* 333:742.

4. Nemazee, D.A. and K. Bürki. 1989. Clonal deletion of B lymphocytes in a transgenic mouse bearing anti-MHC class I antibody genes. *Nature* 337:562.

5. Hartley, S., J. Crosbie, R. Brink, A. Kantor, A. Basten, and C. Goodnow. 1991. Elimination from peripheral tissues of self-reactive B lymphocytes recognizing membrane-bound antigens. *Nature* 353:765.

6. Rudensky, A., S. Rath, P. Preston-Hurlburt, D. Murphy, and C.A. Janeway Jr. 1991. On the complexity of self. *Nature* 353:660.

7. Janeway, C.A., Jr. 1989. Approaching the asymptote? Evolution and revolution in immunology. *Cold Spring Harbor Symp Quant Biol* 54:1.

8. Lanzavecchia, A. 1990. Receptor-mediated antigen uptake and its effect on antigen presentation to class II-restricted T lymphocytes. *Annu Rev Immunol* 8:773.

9. Springer, T.A. 1990. Adhesion receptors of the immune system. *Nature* 346:425.

10. Freeman, G.J., G.S. Gray, C.D. Gimmi, D.B. Lombard, L.-J. Zhou, M. White, J.D. Fingeroth, J.G. Gribben, and L.M. Nadler. 1991. Structure, expression, and T cell costimulatory activity of the murine homologue of the human B lymphocyte activation antigen B7. *J Exp Med* 174:625.

11. Linsley, P.S., W. Brady, L. Grosmaire, A. Aruffo, N.K. Damle, and J.A. Ledbetter. 1991. Binding of the B cell activation antigen B7 to CD28 costimulates T cell proliferation and interleukin 2 mRNA accumulation. *J Exp Med* 173:721.

12. Beverley, P.C.L. 1990. Human T cell memory. *Curr Top Microbiol Immunol* 159:111.

13. Hayakawa, K. and R. Hardy. 1991. Murine CD4[+] subsets. *Immunol Revs* 123:145.

14. Pfeiffer, C., J. Murray, J. Madri, and K. Bottomly. 1991. Selective activation of Th1- and Th2-like cells in vivo-- resonse to humna collagen IV. *Immunol Rev* 123:65.

15. Abbas, A., M. Williams, H. Burstein, T.-L. Chang, P. Bossu, and A. Lichtman. 1991. Activation and functions of CD4[+] T cell subsets. *Immunol Revs* 123:5.

16. Metlay, J.P., E. Puré, and R.M. Steinman. 1989. Control of the immune response at the level of antigen-presenting cells: a comparison of the function of dendritic cells and B lymphocytes. *Adv Immunol* 47:45.

17. Bretscher, P. and M. Cohn. 1970. A theory of self-nonself discrimination. *Science* 169:1042.

18. Lafferty, K.J., S.J. Prowse, C.J. Simeonovic, and H.S. Warren. 1983. Immunobiology of tissue transplantation: a return to the passenger leukocyte concept. *Annu Rev Immunol* 1:143.

19. Schwartz, R.H. 1990. A cell culture model for T lymphocyte clonal anergy. *Science* 248:1349.

20. DeSilva, D., K. Urdahl, and M. Jenkins. 1991. Clonal anergy is induced in vitro by T cell receptor occupancy in the absence of proliferation. *J Immunol* 147:3261.

21. Rammensee, H.-G., R. Kroschewski, and B. Frangoulis. 1989. Clonal anergy induced in mature Vβ6[+] T lymphocytes on

immunizing Mls-1^b mice with Mls-1^a expressing cells. *Nature* 339:541.

22. Rocha, B. and H. von Boehmer. 1991. Peripheral selection of the T cell repertoire. *Science* 251:1225.

23. Webb, S., C. Morris, and J. Sprent. 1990. Extrathymic tolerance of mature T cells: Clonal elimination as a consequence of immunity. Cell 63:1249.

24. Lo, D., L.C. Burkly, R.A. Flavell, R.D. Palmiter, and R.L. Brinster. 1990. Antigen presentation in MHC II transgenic mice: Stimulation versus tolerization. *Immunol Rev* 117:121.

25. Blackman, M., T. Finkel, J. Kappler, J. Cambier, and P. Marrack. 1991. Altered antigen receptor signaling in anergic T cells from self-tolerant T-cell receptor β-chain transgenic mice. *Proc Natl Acad Sci USA* 88:6682.

26. Sprent, J., E.-K. Gao, and S.R. Webb. 1990. T cell reactivity to MHC molecules. *Science* 248:1357.

27. Marrack, P. and J. Kappler. 1990. The staphylococcal enterotoxins and their relatives. *Science* 248:705.

28. Blackman, M., J. Kappler, and P. Marrack. 1990. The role of the T cell receptor in positive and negative selection of developing T cells. *Science* 248:1335.

29. Möller, G., ed. *Transgenic mice and immunological tolerance.* Immunol. Revs., Vol. 122. 1991.

30. Green, D., P. Flood, and R. Gerson. 1983. Immunoregulatory T-cell pathways. *Annu Rev Immunol* 1:439.

31. Powrie, F. and D. Mason. 1990. OX-22^{high} CD4$^+$ T cells induce wasting disease with multiple organ pathology: prevention by the OX-22^{low} subset. *J Exp Med* 172:1701.

32. Coutinho, A. and A. Bandeira. 1989. Tolerize one, tolerize them all: tolerance is self-assertion. *Immunol Today* 10:264.

33. Lo, D., L.C. Burkly, R.A. Flavell, R.D. Palmiter, and R.L. Brinster. 1989. Tolerance in transgenic mice expressing class II major histocompatibility complex on pancreatic acinar cells. *J Exp Med* 170:87.

34. Parker, D.C. and E.E. Eynon. 1991. Antigen presentation in acquired immunological tolerance. *FASEB J* 5:2777.

35. Gosselin, E.J., H.-P. Tony, and D.C. Parker, Normal resting B cells process antigen, in *Antigen Presenting Cells: Diversity, Differentiation, and Regulation,* L.B. School and J.G. Tew, Editor. 1988, A.R. Liss, Inc.: New York. p. 341.

36. Sprent, J. 1980. Features of cells controlling H-2 restricted presentation of antigen to T helper cells in vivo. *J Immunol* 125:2089.

37. Lassila, O., O. Vaino, and P. Matzinger. 1988. Can B cells turn on virgin T cells? *Nature* 334:253.

38. Hori, S., S. Sato, S. Kitigawa, T. Azuma, S. Kokudo, T. Hamaoka, and H. Fujiwara. 1989. Tolerance induction of allo-class II H-2 antigen-reactive L3T4+ helper T cells and prolonged survival of the corresponding class II H-2-disparate skin graft. *J Immunol* 143:1447.

39. Tony, H.-P. and D.C. Parker. 1985. Major histocompatibility complex-restricted, polyclonal B cell responses resulting from helper T cell recognition of antiimmunoglobulin presented by small B lymphocytes. *J Exp Med* 161:223.

40. Mitchison, N.A. 1964. Induction of immunological paralysis in two zones of dosage. *Proc Roy Soc Ser Biol Sci* 161:275.

41. Eynon, E.E. and D.C. Parker. 1992. Small B cells as antigen presenting cells in the induction of tolerance to soluble protein antigens. *J Exp Med* 175:in press.

42. Yuschenkoff, V.N., E. Eynon, and D.C. Parker. 1991. Tolerance induction to a protein antigen using B cells from a transgenic mouse. *J Cell Biochem* Suppl. 15A:245.

43. Finkelman, F.D., I. Scher, J.J. Mond, S. Kessler, J.T. Kung, and E.S. Metcalf. 1982. Polyclonal activation of the murine immune system by an antibody to IgD. II. Generation of polyclonal antibody production and cells with surface IgG. *J Immunol* 129:638.

44. Stockinger, B. and B. Hausmann. 1988. Induction of an immune response to a self antigen. *Eur J Immunol* 18:249.

45. Ashwell, J.D. 1988. Are B lymphocytes the principal antigen presenting cells in vivo? *J Immunol* 140:3697.

46. Lin, R.-H., M.J. Mamula, J.A. Hardin, and C.A. Janeway. 1991. Induction of autoreactive B cells allows priming of autoreactive T cells. *J Exp Med* 173:1433.

7

THE FUNCTIONAL RELATIONSHIP BETWEEN PREFERENTIAL USE OF NEWLY SYNTHESIZED CLASS II MOLECULES AND THEIR STABLE ASSOCIATION WITH PEPTIDES

Colin Watts[*] and Antonio Lanzavecchia[#]

* Department of Biochemistry, University of Dundee, U.K.
\# Basel Institute for Immunology, Basel, Switzerland

SUMMARY

Pre-existing class II molecules cannot present newly processed peptides (i) because they do not reenter the processing compartment and (ii) because they are already involved in an essentially irreversible association with peptide. Class II biosynthesis and not simply surface class II expression is the hallmark of an effective APC.

NATURALLY PROCESSED ANTIGEN BINDS TO NEWLY SYNTHESIZED CLASS II MOLECULES

An increasing body of evidence indicates that class II MHC molecules acquire processed peptides on the biosynthetic route (1-3). Once on the cell surface it is not clear how longlived these complexes are and it has often been suggested that, in vivo, MHC class II molecules may be reused (4,5).
While peptide exchange would offer increased capacity for binding incoming antigens it would inevitably reduce the efficacy of processed antigenic peptides by exchanging these for self peptides.
Using a direct biochemical readout for loading of class II molecules with naturally processed antigen we have recently found that there is a marked preference for the binding of iodinated antigen fragments to the newly synthesised class II molecules (3). In the same experiments exogenously added peptides bound preferentially to surface class II molecules. Thus, while surface class II molecules have some binding capacity for peptides and although they enter the endocytic pathway (6), the rapid recycling kinetics appears to preclude access to the antigen processing compartment.

Thus, these results cast some doubt on the relevance of an exchange mechanism for antigen presentation. Indeed, if newly processed peptides can only bind to new class II molecules, peptide displacement/exchange would reduce the efficacy of antigen presentation without offering more capacity for binding of newly processed peptides.

Since surface Class II molecules are not reused to bind incoming antigens then one would predict that the peptide bound during maturation should remain bound, perhaps over the lifetime of the molecule, otherwise a high proportion of empty molecules would be found.

CLASS II MOLECULES BIND PEPTIDES IRREVERSIBLY IN LIVING CELLS

Earlier results showed that in some cases antigen-pulsed specific B cells could stimulate T cells up to 12 days after the initial exposure (7). Since the biochemical halflife of intact antigen in these B cells was only 2-3 hours (8) this suggested that this persistence was not due to a very long lived antigen store but instead to very stable peptide/MHC complexes.

We have measured the halflife of peptide class II complexes using both a functional and biochemical assay (9). To measure the functional halflife of peptide/MHC complexes we pulsed various types of antigen presenting cell (APC) with different concentrations of synthetic peptides corresponding to known T cell epitopes. The cells were recultured in the absence of peptide and at different time points the level of stimulatory complexes was measured by titrating the number of APC in a T cell proliferation assay (10). By plotting the number of APC required for a given level of T cell stimulation verus time one obtains a measure of the half life of the complexes which is independent of the level of peptide used and is ~ 25 hours on EBV B-cells. A similar halflife was estimated on activated T cells and peripheral blood mononuclear cells (PBMC) using four different peptide epitopes.

An independent and direct estimate of complex stability was obtained by measuring the lifetime of radioiodinated peptide/MHC complexes. APC were pulsed with radioiodinated peptides. After different times of chase at 37°C we immunoprecipitated class II DR molecules and measured the labelled peptide bound. The lifetime of these complexes was in good agreement with that indicated by functional assays, i.e. 30-40 hrs.

Peptides were able to compete with each other for binding to a given class II allele. However, they were not able to displace iodinated peptide that were already bound to class II molecules.

Thus both functional and biochemical assays using different epitopes and different APC indicate that peptide/MHC class II complexes are very stable in living cells with a suprisingly reproducible halflife of ~30 hours. This suggested that for these epitopes, persistence might be determined by the lifetime of the class II molecules themselves. Indeed, in agreeement with earlier reports we found that the halflife of class II molecules in our EBV-B cells is ~30 hrs. We conclude that immunogenic peptides are bound essentially irreversibly to class II molecules, are not displaced under physiological conditions and their turnover reflects the turnover of class II molecules themselves.

These epitopes represent a minor fraction of the class II molecules and were selected because of their immunogenicity. Thus their behavior may not reflect the bulk of the class II bound peptides. Using a novel technique based on diagonal SDS/SDS urea gels we were able to display low Mr peptides from disrupted class II $\alpha\beta$ complexes (9). The results show that the halflife of this low molecular weight material was strikingly similar to that of the α and β chains from which it was eluted, i.e. ~30 hours on EBV-B cells.

In summary, using both functional and biochemical readouts on three different cell types we measured the halflife of the complexes of DR with three synthetic peptides as well as with presumptive endogenous peptides and found that class II molecules can bind peptides irreversibly in the sense that the halflife of the complex is determined by the halflife of the class II molecules themselves (9).

LIFE HISTORY OF CLASS II MOLECULES IN THE CONTEXT OF ANTIGEN PRESENTATION

The genesis and maturation of class II molecules is now relatively well undestood and has beeen recently reviewed (11), but in terms of its function for antigen presentation we should consider its overall lifetime. As discussed above, there is now good evidence that peptides are acquired on the biosynthetic route and retained for the life of the class II molecules. As it turns out, in human B cells recycling of class II seems to be of limited functional importance, since it does not result in peptide exchange (at least on the majority of the molecules).

An important working principle emerges from these results: an APC for class II is best defined not as a cell with high levels of surface class II molecules, but as one that is actually synthesizing these molecules. This is perhaps why class II biosynthesis is tightly regulated in APC such as B cells and macrophages. Following appropriate stimulation of synthesis by lymphokines increased numbers of MHC molecules will be synthesized and made available for binding incoming antigens. The preexisting MHC molecules on the cell surface may indeed not only be useless but even play a negative role by "distracting" the CD4 molecules or even th T cell receptor from the productive interaction. In fact, the best APC may be those expressing a high specific peptide density irrespective of the total level of class II molecules.

We suggest that the persistence of peptide/MHC complexes is a critical factor for the immune response since it allows the APC to keep the 'memory' of antigen encountered over a short time window allowing presentation at a distant site perhaps several days later. The combination of inducible class II MHC synthesis and persistent complexes will optimise presentation of antigen and is perhaps best illustrated by dendritic cells isolated from antigen pulsed animals which retain the ability to trigger T cells specific for that antigen but fail to process and present a second antigen added during in vitro culture (12). Thus the life history of most class II molecules is characterised by a long monogamous association with a peptide encountered early in its lifetime.

ACKNOWLEDGEMENTS: We thank Stephane Demotz and Paolo Dellabona for reading the manuscript. This work was partially supported by the Wellcome Trust and the Medical Research Council. The Basel Institute for Immunology was founded and is supported by F. Hoffmann La Roche & Co. Ltd., Basel, Switzerland.

REFERENCES

1.   Adorini, L., Ullrich, S.J., Apella, E. &Fuchs, S. *Nature* **346**, 63-67 (1990).
2.   Peters, P.J., Neefjes, J.J., Oorschot, V., Ploegh, H.L., & Geuze, H.J. *Nature* **3 49**, 669-676 (1991).
3.   Davidson, H.W., Reid, P.A., Lanzavecchia, A. & Watts, C. *Cell* **67**, 105-116 (1991).
4.   Harding, C.W., Roof, R.W., & Unanue, E.R. *Proc. Natl. Acad. Sci. USA* **86**, 4230-4234 (1989).
5.   Adorini, L., Appella, E., Doria, G., Cardinaux, F., & Nagy, Z.A. *Nature* **342**, 800-803 (1990).
6.   Reed, P.A. & Watts, C. *Nature* **346**, 655-657 (1990).
7.   Lanzavecchia A, *Immunol. Rev.* **99**, 39-51 (1987).
8.   Watts, C. & Davidson, H.W. *EMBO J.* **7**, 1937-1945 (1988).
9.   Lanzavecchia, A., Reid, P.A. & Watts, C. *Nature*. 1992. (in press).
10.   Mathis L.A., Glimcher, L.H., Paul, W.E., & Schwartz, R. H. *Proc. Natl. Acad. Sci. USA* **80**, 6019-23 (1983).
11.   Neefjes, J.J. & Ploegh, H.L. 1992. *Immunol.Today* **13**: 179.
12.   Pure`, E., Inaba, K., Crowley, M.T., Tardelli, L., Witmer -Pack, M.D., Ruberti, G., Fathman, G. & Steinman, R.M. *J. Exp. Med.* **172**, 1459-1469 (1990).

8

EPITOPE SELECTION AND AUTOIMMUNITY

N.A. Mitchison

Deutsches Rheuma-Forschungszentrum Berlin
Am Kleinen Wannsee 5
D-1000 Berlin 39

It is now well established that T cells normally
recognize only a limited number out of the total universe of
foreign epitopes potentially available in an invading
organism, and that likewise they are tolerant through
negative selection of only a limited number of self-
epitopes. This principle is illustrated in figure 1, where
just two epitopes on a protein are depicted as being
recognised. Many proteins, both foreign and self, are
probably not normally recognised at all. The principle has
many important and interesting ramifications. In terms of
mechanism, current research is beginning to dissect out the
steps by which epitopes are selected for recognition. The
evolutionary pressures which control the selection process
are beginning to be understood. They range from the ever
ongoing struggle between host and parasite, to the need to
balance preservation of the T cell repertoire against the
threat of autoimmunity. There are important implications for
vaccine design, where the possibility of bringing hitherto
unrecognised epitopes into play motivates much of current
research on the molecular biology of parasites. There are
implications also in autoimmunity. The narrower the scope of
negative selection, the greater the danger of autoimmune
disease. The purpose of this article is to explore these
ramifications.

HISTORICAL BACKGROUND

In its original form, as proposed by Burnet and
analysed by Billingham, Brent and Medawar (1), tolerance of
self was thought to apply to most self-macromolecules. In
the nomenclature which is used today, such molecules were
believed to mediate negative selection, as a consequence of
which T cells potentially able to react with self are
missing from the mature T cell repertoire. Yet even then it
was already suspected that not all self-macromolecules would
behave in this way, because some of them would be confined

Figure 1. T cells recognise only limited parts of proteins

to sites that had no access to the immune system. A few
years earlier Medawar (2) had discovered that allografts
transplanted into the brain or anterior chamber of the eye
do not get rejected, and had drawn the conclusion that these
sites enjoyed this "privilege" because of their anatomical
isolation from cells of the immune system. Eventually one
came to accept a proportion of self-macromolecules as
"immunologically silent" because of sequestration within
anatomically isolated compartments, or because of
confinement within the interior of cells, or simply from
their very low concentration in body fluids (3).

The scope of negative selection was also narrowing in
other ways. It became clear that B cells obey different
rules from T cells, being less tightly circumscribed by
negative selection. As so often happens in the history of
science, what starts as an empirical generalisation ends up
as a logical necessity. In this particular case it was
discovered - much later - that B cells hypermutate. From the
point of view of maintaining self-tolerance through negative
selection this behavior can be regarded as irresponsible,
and this logically explains why so many epitopes
recognisable by B cells survive on self-proteins. Indeed it
is puzzling why B cells become tolerant at all, and this is
a topic to which we shall return.

In spite of this narrowing, the scope of negative
selection still seemed quite extensive. Indeed once it
became clear that T cells could respond to just a short
peptide bound to an MHC molecule, one began to wonder how
any of them could escape negative selection. The next step
forward was to examine in detail the peptides which actually
did get recognised. So far as the CD4 (MHC class II
restricted) T cells are concerned, regulatory cells which
are of most interest in the present context, we are largely
dependent on the work of Sercarz and his colleagues. They
analysed the way in which mouse T cells recognise such
foreign proteins as avian lysozymes and bacterial
galactosidase. As Sercarz's own contribution here describes,
these proteins were cleaved into their constituent peptides,
or synthesised as sets of smaller peptides. When presented
to T cells primed with the intact protein, a common finding
was that surprisingly few of these peptides prove able to
stimulate. Evidently the intact protein delivered only a
small number of effective T cell epitopes. In contrast, many

more of the peptides prove able to induce a T cell response when they are used for immunization on their own.

Selectivity of the same sort operates for cell-endogenous proteins, typically viral proteins, which get cleaved, bound by MHC Class I molecules, and recognized by CD8 cytotoxic T cells. These epitopes seem to be even more stringently selected, as is discussed here in Nathenson's contribution.

The most recent step forward has been to discover that the same kind of epitope selection affects negative selection. That was to be expected, as positive and negative selection must surely keep pace with one another. In an elegant recent study Cibotti et al made transgenic mice that expressed varying amounts of hen egg lysozyme, and found that low levels deleted reactivity to only one dominant epitope in the protein, while higher levels began to impair reactivity to other epitopes (4).

STAGES IN EPITOPE SELECTION

Figure 2 presents a schematic view of the steps leading to stimulation of CD4 T cells, and indicates the stages at which selection could occur. To start with, anatomical barriers or the cell membrane, as mentioned above, and possibly other barriers might bar access to the immune system. Next, portions and perhaps entire proteins might be lost during the processing steps within an antigen presenting cell (a B cell or dendritic cell) prior to presentation. Included here is the possibility of epitope capture by competing MHC molecules, as envisaged by Sercarz and by Nepom (5). The third stage, of presentation, looses the many peptides which do not fit into the MHC molecule's groove. So remarkably selective is this process that one tends to forget the other stages. Then finally, potentially valid presentation could fail because no T cells are there to respond. In the case of a foreign antigen this might result from deletion of the corresponding T cells by negative selection, but other possibilities are open, such lack of peripheral positive selection or conformational problems with the T cell receptor.

It is natural to inquire which of these stages is in practice the more important, but we have to face the fact that there may be no general answer. Certainly the whole process must be tightly subject to natural selection, most notably from parasites (using the term in its widest sense to include everything down to viruses) varying the structure of their more vulnerable proteins so as to escape immunological attack. Less compellingly, one supposes that self-proteins are subject to natural selection so as to minimise the number of epitopes they present, which would in turn minimise the amount of negative selection needed and so preserve as much as possible of the T cell receptor repertoire. A more important factor for self-proteins, probably, is accurate adjustment of negative and peripheral positive selection so as to match one another, and that is our next topic. In the meanwhile it is worth emphasising that there is no *deus ex machina*. Contrary to what biochemists may believe, epitope selection is not imposed

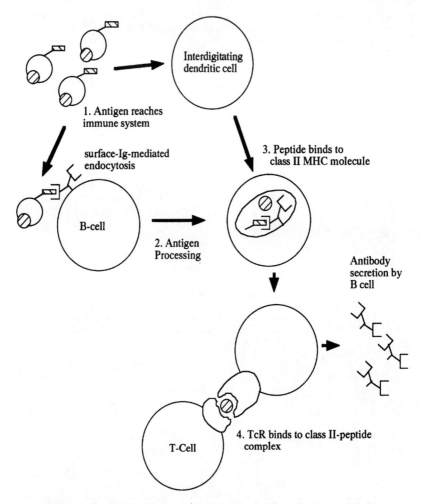

Figure 2. Four stages in T cell activation at which epitope selection can occur

simply by the conformation of MHC molecules. Rather, their confirmation must be delicately controlled by evolutionary pressures. One could hope that an antigen such as hen egg lysozyme would be totally neutral, being neither a parasite nor a self protein, but how sure can one be that the immune system is entirely innocent of exposure to homologous sequences?

Leaving aside the antigen processing and presentation aspects of epitope selection here for Mathis and Sercarz, let us move on to selection via the T cell receptor repertoire.

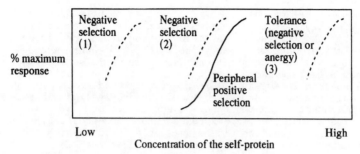

% maximum response

Low   Concentration of the self-protein   High

Figure 3.The balance between negative and peripheral positive selection

THE PRECISE ADJUSTMENT OF NEGATIVE AND PERIPHERAL POSITIVE SELECTION

In this discussion negative selection is used in its usual sense to denote the process of acquisition of tolerance-of-self by developing T cells within the thymus. Peripheral positive selection refers to the response of mature T cells to antigen which they encounter in the periphery (rather than to positive selection as it occurs in the thymus prior to negative selection). Figure 3 depicts various possible relationships between these two modes of selection. The curve of peripheral positive selection shown in the figure represents simply the familiar proliferative response of primed T cells, plotted as a function of antigen concentration. Curves of this sort have been obtained for many soluble proteins and peptides, all of which show much the same concentration dependence. Precise measurements of the concentrations needed for negative selection have been harder to obtain, and I shall describe a new system that we have been using for this purpose (6). Before doing so, consider the three possibilities set out in the figure as they would apply to a self-protein. Note that only the top half of the curve is depicted, because complete data covering the weaker tolerance effects are hard to collect. At position (1), the negative selection curve is placed far to the left. The effect would be to generate mature T cells that would be safe but ineffective, like the Bundeswehr. They could never generate an autoimmune response against the self protein in question, but on the other hand would lack large parts of the T cell receptor repertoire. Position 2 looks as though it would provide an optimal balance between these two opposing needs. Position 3, with the negative selection curve placed far to the right, would be a disaster, as it would leave potentially self-reactive T cells. Even so, as I shall argue below, this last possibility does seem to be the one which applies to B cells.

It may come as no surprise to learn that the system we chose for our study was the response of mice to F liver protein, as we have been publishing on this subject over the last twenty years. Indeed we undertook the study precisely because this system seemed so well adapted to the purpose.

What was needed was (1) a well-defined soluble protein, preferably a self-protein which normally induces tolerance within the thymus (so as to model the natural processes of selection), and preferably one which is present in body fluids at low concentrations (so as to minimise interference from B cells effects). On the other hand the protein should be reasonably accessible in quantity. In addition (2), it should not be be expressed within the thymus, as otherwise its concentration could not be accurately controlled(this excludes most proteins expressed off transgenes). Finally, (3) a good method should be available for detecting its primary T cells response. The F protein system met all these requirements.

Briefly, fetal thymus lobes were cultured *in vitro* under conditions in which they generated substantial numbers of mature CD4 T cells. These cells were then transferred into irradiated syngeneic hosts, where they were immunised with self-F protein in adjuvant. Their response to this challenge was measured by transferring in primed B cells plus further self-F protein (without adjuvant), and reading out the ensuing anti-F antibody response. We could then begin adding F protein into the culture system, so as to find out where the negative selection curve is placed. We could compare it with the curve for peripheral positive selection, which had been obtained by conventional methods. We discovered that the two curves in fact bore the relationship indicated by position 2 in figure 3. So pleased were we with these data that we named the two curves "The west wind blows a cloud trail from Mt. Fuji", because the dotted curve does look something like a cloud.

The implication is that negative and peripheral positive selection are exquisitely balanced, so as to preserve as much of the repertoire as possible without paying too much of a penalty in terms of autoimmunity. That is not unexpected. I wish to draw attention to the extreme importance of the partitioning between self and non-self ("the relative size of self") discovered by Seiden and Celada in their model of the immune system, which makes essentially the same point. This concurrence, which for me is the major message from Sardinia, was arrived at quite independently as we neither of us knew of one another's work on the subject.

Before closing discussion of this experiment, it is worth summarising how far it takes us in understanding the mechanism of negative selection. The conclusion here of von Böhmer's is that negative selection occurs among CD4CD8 double positive thymocytes, and this resurrects the old Lederburg hypothesis that tolerance-susceptibility is stage specific. All the evidence comes from experiments performed with membrane-bound antigens. In contrast the experiment described above reveals a form of thymic negative selection which is more or less identical, in terms of antigen concentration required, to that obtainable in mature (peripheral) T cells (7). This resurrects another old hypothesis, that negative selection depends on the lack of some co-stimulus. In the old days this was formulated as a lack of "macrophage presentation" (8). These days one looks for some defined molecular interaction with antigen presenting cells inducing apoptosis. To judge from unpublished data recently presented by Dr Craig Thompson, binding of CD28 to BB7 is an promising candidate for this role. We are left with the disconcerting possibility that soluble self-proteins operate negative selection on principles different from those of membrane-bound ones.

A second point can be made, also related to heterogeneity of thymocytes. Like mature T cells, CD4CD8 double positive cells comprise a mixture of CD45RO and CD45RA cells, although they include few of the latter. In the periphery CD45RO is a marker of activation, rather than of memory as had at one time been thought (9), and cells with this phenotype are hyper-responsive to antigen. In terms of the balance that we have been discussing, it makes sense for negative selection to occur among cells that are at their most responsive, as otherwise non-specific activators (superantigens etc) would turn on autoimmunity. Indeed a colleague had already reached the conclusion, based on some rather contorted mathematics, that there simply are not enough CD45RA cells in the thymus for them to constitute the population that gets negatively selected (10). That leaves us with a kaleidoscopic view of T cell development, in which we start with a CD45R⁻ progenitor, move on to a CD45RO stage that is subject to negative (and intrathymic positive?) selection, then comes to rest as a CD45RA T cell in the periphery, can then be stimulated to become CD45RO again during peripheral positive selection, and would then revert to CD45RA as a long-term memory cell.

POSITIVE SELECTION OF THE PERIPHERAL REPERTOIRE

Having completed our discussion of negative selection and its impact on epitope selection, we need to consider the other side of the coin, the impact of peripheral positive selection. This is an important aspect of T cell biology, but we can be brief because only the surface of this subject has yet been scratched. The technologies discussed here by Karjalainen (expression of soluble T cell receptors and assay of their function) and Theofilopoulos (V gene usage analysis) have much to contribute. In the meanwhile, we look with envy at the technologies which the B cell biologists are bringing to bear: cloning expressed V genes out of single cells, identifying memory cells by the presence of hypermutated segments, and examining V gene function by transfection. In comparison, problems with T cells seem dauntingly difficult.

One controversial point concerns Cohen's argument that the mature T cell repertoire is strongly selected by such self-proteins as heat shock proteins (HSP's) and myelin basic protein. The high level of reactivity that is evident towards bacterial HSP's can indeed be interpreted in this way, but there is another possibility. Environmental bacteria have a huge impact on T cells, as demonstrated by the lack of such cells in germ-free animals. Highly conserved proteins such as the HSP's share epitopes among different bacterial species, and can therefore be expected to dominate a mixed anti-bacterial response. Let me then put on record the recent finding (made with funding from the

Rheuma Center) that human cord blood T cells already show a high level of reactivity to one of the HSP's, i.e. prior to exposure to environmental bacteria (11). One up to Irun Cohen!

TOLERANCE AMONG B CELLS

This subject has a long and vexed history. Early work established that foreign proteins administered by repeated injection could much more easily induce tolerance among T than B cells (12). Later it was discovered that freshly minted B cells, just before they acquire surface IgD, transiently become highly sensitive to tolerance induction, and so it was argued that self-proteins might induce tolerance at a lower concentration threshold than had been thought. This seemed to me unrealistic, because most of the foreign proteins studied were serum proteins which take some time to clear, and also because it doesn't make much sense to tolerise B cells without involving the T-B interaction. At any rate, a list of extracellular self-proteins began to accumulate where tolerance could be detected only among T cells, such as C5, F liver protein, and thyroglobulin (13). This does not apply to high concentration serum proteins, where B cells do become self-tolerant. For an elegant example, see the the absence of antibodies reactive with self immunoglobulin allotypes among rheumatoid factors (14). Meanwhile self-membrane-proteins were found to behave rather differently, since nearly all of those tested did induce tolerance among B cells. The evidence came generally from allo-immunisation, where usually there is no difficulty in raising antibodies specific for allotypic variants without cross-reacting with host cells. Admittedly there are a few exceptions, such as the transient Coombs-positivity which develops on host erythrocytes when mice are immunised with rat red blood cells. As for intracellular proteins, I know of no examples of their inducing tolerance among B cells, with the often quoted exception of haemoglobin based on the work of Reichlin (discussed in 12). And anyone who has watched a bruise knows how easily haemoglobin leaks.

This information is summarised in figure 4. One might suppose that soluble and membrane-bound proteins operate negative selection in different ways for purely mechanical ways, for instance through their interactions with the surface immunoglobulin receptors of B cells. It is more appealing to think adaptively, taking account of the relative hazards posed by different anti-self responses as indicated in the figure. Presumably anti-cell antibodies must be fairly dangerous, even though the transient Coombs positivity referred to above seems to do the mice no harm. Antibodies to extracellular proteins are often surprisingly well tolerated. Most of the hundred or so Indian ladies who have been given the Talwar birth control vaccine make antibodies reactive with their own luteinizing hormone, but do not even suffer interruption of the menstrual cycle (15). Once again, natural selection seems to deliver only what is needed.

It is tempting to press the evolutionary argument even further. Maybe an important function of B cell tolerance is

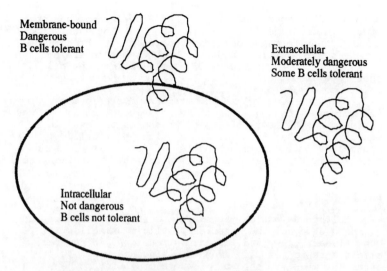

Membrane-bound
Dangerous
B cells tolerant

Extracellular
Moderately dangerous
Some B cells tolerant

Intracellular
Not dangerous
B cells not tolerant

Figure 4. Autommunity to some self-proteins
is more dangerous than to others

to protect membrane proteins against a possibility evident
in figure 3. Maybe (a lot of maybes in this paragraph!)
antigen-specific B cells present certain epitopes to T cells
additional to those presented constitutively by dendritic
cells. This possibility is inherent in the concept of
vectored presentation, discussed here by Manca (although his
own data do not directly support this possibility, so far as
I can see). The idea is that the B cell's antibody receptor
might protect certain sequences of the ingested antigen
against cleavage. Dr Herman Waldmann has given me a
beautiful example of this possibility from his group's work
on anti-cell antibodies (ACA). They sought to prevent HAMA
(host anti-mouse antibody) responses to mouse monoclonal ACA
by prior tolerisation with non-cell-binding mouse Ig. The
tolerisation was carried out with ACA H chain bound to
third-party L chain mixed with ACA L chain bound to third-
party H chain. Although they did obtain tolerance of the
mixture, the animals could still respond to the intact ACA,
indicating that the foreign immunoglobulin presented new
determinants when processed via binding to host cells.

If there is anything in this possibility, negative
selection mediated only by the dendritic cells of the thymus
could be dangerous. The immune system could avoid the danger
by tolerising B cells, and that would be all the more
important for self-proteins bound to cell membranes. In his
contribution here Parker raises the interesting possibility
that B cells may themselves mediate peripheral tolerance,
but in the thymus there are too few of them for much to be
expected (discussed in 6).

Our discussion has used the term "tolerance" in
connection with B cells. That is because we know that these
cells do not respond, but not whether they have been

eliminated or anergised. Anergised B cells crop up in studies with transgenes, as discussed here by Mathis, but whether they form a significant component of the normal immune system remains in doubt. What is needed are good markers of the anergised cell, a well-defined anergised phenotype, and until that is available the question is likely to remain open.

Although B cell tolerance remains mysterious, as do the immunological consequences of being membrane-bound, we can begin to see glimmers of light. This area of research is likely to have a rich future, in which genetically manipulated mice lacking B cells (16) and self-proteins genetically manipulated so as to alter their location should prove gold mines.

LATENT HELP

Epitope selection may explain a surprise encountered during a study of the kinetics of alloimmunisation (17). Most alloantigens, and indeed most antigens in general, elicit a rapid response among T cells. Reaching a plateau after one week seems to be a general rule for mouse immunization. Certain antigens however, such as allo-Thy1 and allo-MHC class I, take very much longer as illustrated in figure 5. Just why that should be so is completely unclear, but it is tempting to speculate that the same mechanisms which protect membrane-bound self-proteins may be at work. I refer to this phenomenon as "latent" help, simply because this response on the part of helper T cells is not disclosed by conventional short-term immunisation. This terminology overlaps that of Sercarz, who prefers the term "cryptic", and for the time being they can both be used co-extensively.

THE VACCINATOR'S DILEMMA

Epitope selection has important implications for vaccine design. A bird's eye view of the last decade would show it starting in a flurry, as parasitology woke up under the impact of molecular biology, and the possibility of second generation vaccines based on genetic engineering became evident. Yet the outcome up to the present has been somewhat disappointing. We have come to realise that parasites and hosts have been negotiating with one another over millions of years, leaving equilibria that are not easy to disturb. Nowhere is that more evident than in the failure so far of the world-wide effort to develop a malaria sporozoite vaccine.

More than anything else, perhaps, the effort to develop new vaccines has been inspired by the idea that finding just the right epitopes may let us perturb the equilibria. The problem is that we do not know quite what to look for. Are we hunting for immunodominant epitopes, which would generate a tremendous response if only they could freed of the suppressor epitopes which normally accompany them? Or are we hunting for latent ones, potentially able to give the

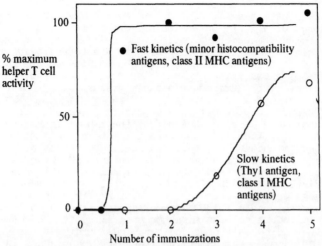

Figure 5. Most antigens induce helper T cell activity rapidly,
but some do not.
These data were collected in adoptive transfers,
in which T cells recognized the antigens specified,
while B cells recognised Thy1 or class I MHC antigens.

parasite a nasty surprise of a kind that it has never before
encountered?

My impression is that immuno-parasitology is still
full of vim and vigour, and that these questions will get
resolved by trial and error. If I do have a personal
opinion, it is that perhaps more effort should go into
collecting, enumerating, and classifying epitopes, and less
into promoting particular vaccines at a premature stage. The
natural history of epitopes is a worthwhile subject for
study.

THE GENESIS OF AUTOIMMUNITY

Opinion on this subject is broadly divided, according
to whether one regards as more important mistakes in the
thymus or in the periphery. Here von Böhmer espouses the
former view, with his proposal that cells which express the
T cell receptor before expressing CD4 and CD8 may be the
culprits. I prefer the latter, but before going further let
me emphasise that in no case is the initiation of an
autoimmune disease properly understood. We are not short of
speculation about triggering events, where attention has
focussed on fairly common viral infections. Most physicians,
I suspect, would place their bets in that area. But the kind
of prospective epidemiological study that might settle the
question has yet to be undertaken: the scale and expense
would need to be enormous.

My basic reason for attaching more importance to the
periphery is that epitope selection makes every self-protein

a time bomb. That view has become more compelling as the evidence of epitope selection has grown. It has become increasingly clear that most or all self-proteins contain "cryptic" or "latent" epitopes, that are potentially able to generate autoimmune disease if accessed. Indeed the problem is the plethora rather than the shortage of latency. Why do the majority of autoimmune disease attack only synovia, oligodendrocytes, or islets?

Figure 5 above offers a suggestive model for the onset of autoimmune disease. Neither of the two antigens that display slow kinetics, thy1 and allo-MHC class I, are much good as immunogens on their own (allo-MHC refers here only to naturally occurring molecules; mutant MHC antigens behave differently, as I have discussed elsewhere (17)). The kinetics shown in the figure were obtained only when these antigens were presented in conjunction with other, better immunogens; the allo-MHC molecules presented on their own tend to induce tolerance rather than immunity. One can easily imagine self-proteins behaving in this way as "latent" immunogens, that require help from viruses in order to provoke an immune response.

Our Rheuma Center has begun to study one particular example of autoimmunity in the field of rheumatic diseases. Lupus, Sjögren's disease, and certain other connective tissue diseases have characteristic constellations of "marker" antibodies. Patients develop a group of antibodies, all of which recognise a single type of cellular particle, such as nuclear RNA-binding particles or microtubule organising centres. What happens, we believe, is that T cells come to recognise one or more latent epitopes in the particle, and then bring groups of B cells into the response. In this situation the B cells give one a handle. By cloning them (using EBV transformation), we hope to be able to fish for the T cells that are the real culprits.

If latent epitopes are so dangerous, the immune system could reasonably be expected to have developed secondary defences against them. This is how I see the HLA protective genes that display negative associations with the autoimmune diseases (5, 18). The HLA-DQ genes that are protective in diabetes provide the most convincing examples of this effect. There perhaps lies the solution to the problem of the narrow range of clinically important autoimmunities. Maybe the joints, the central nervous system, and the pancreas are unique not so much in their disease-initiators as in their lack of protectors.

THE SELECTIVE PRESSURES THAT DETERMINE THE LENGTH OF THE PEPTIDE-BINDING GROOVE

As is discussed elsewhere in this volume, an important recent discovery about MHC molecules is that their groove accomodates peptides of sharply defined length: 8-10 amino acids, and for class I molecules usually precisely 9. One might suppose that this would reflect simply the maximum length possible for a molecule of immunoglobulin superfamily type, but this would run contrary to the whole drift of the argument presented in this article. According to this view we should be able to identify the selective pressures acting on the length of the groove, and then test our ideas against a mathematical model.

Reasonable guesses are that the main reason for lengthening groove would be to conserve as much as possible of the T cell repertoire, and for shortening it to minimise the likelihood of parasites avoiding recognition. Thus a shorter groove (accommodating <8 amino acids) would mean that too large a proportion of the possible peptides would be represented within self, and this would engender too much negative selection. A longer groove (>10 amino acids) would too easily let parasites evolve proteins devoid of peptides able to fit.

Modelling these ideas looks to me well within the capabilities of the wonderful Celada-Seiden automaton, although I am not the best judge. Perhaps it would have difficulty coping with the full twenty possible amino acids at each position. At any rate, it is a pleasure for me to issue it this challenge.

REFERENCES

(1) Billingham, R.E., Brent, L., Medawar P.B. 1956.
     Phil. Trans. R. Soc. Lond. B 239:357-414
(2) Medawar, P.B. 1948. Brit. J. exp. Path.29: 58-69
(3) Mitchison, N.A. 1978. Clinics in Rheumatic Diseases
     4: 539-548
(4) Cibotti, R., Kanellopoulos, J.M., Cabaniols, J-
     P.,Halle-Panenko, O., Kosmatopoulos, K., Sercarz,
     E., Kourilsky, P. 1991.Proc. Nat. Aad. Sci. USA.
     In press
(5) Nepom, G.T. 1990. Diabetes, 39:1153-1157.
(6) Robertson K., Schneider, S., Simon, K., Timms E.,
     Mitchison, N.A. 1991. Eur.J. Immunol, in press
(7) Mitchison, N.A. 1968. Immunology 15: 509-530
(8) Dresser, D.W. and Mitchison, N.A. 1968. Adv. Immunol.
     8: 1059-1079
(9) Lightstone, E., Marvel, J., Mitchison, N.A. 1991.
     Eur.J.Immunol, in press
(10) Lightstone, E. 1990. Immunology 71:467-473
(11) Fischer, H, Panayi, G. 1991. Eur. J. Immunol, submitted
(12) Howard, J.G., Mitchison, N.A. 1975. Progr.Allergy
     18: 43-96
(13) Mitchison, N.A. 1985. Clinical Immunology Newsletter
     6: 1, 12-14
(14) Jones, V.E., Puttick, A.H., Williamsonm, E.A.,
     Mageed, R.A. 1988. Clin. exp. Immunol. 71: 451-458
(15) Mitchison, N.A. 1990 Current Opinions in Immunology
     2: 725-727
(16) Kitamura, D., Roes, J., Kühn, R., Rajewsky, K. 1991.
     Nature 350: 423-425
(17) Mitchison, N.A. 1991. Eur. J. Immunol. in press
(18) Baisch, J.M., Weeks, T., Giles, R., Hoover, M., Stastny
     P., Capra J.D., Engle, N. 1990. New England J.Med.
     322: 1836-1844

9

HLA-A2 ALLORECOGNITION AND SUBTYPE DIVERSIFICATION

A. Raúl Castaño and José A. López de Castro

Dpt. of Immunology, Fundación Jiménez Díaz-CSIC
Avda. Reyes Católicos 2, 28040 Madrid, Spain

INTRODUCTION

Analysis of structure-function relationships in HLA class I molecules has traditionally taken advantage of the characterization of subtypes of serologically defined antigens that are differentially recognized by cytotoxic T lymphocytes (CTL). Study of these subtypes has allowed the mapping of amino acid residues in the HLA molecule that determine the specificity of recognition by CTL (1). These functional positions are mainly located in the antigen binding site and alter both self-restricted and allospecific CTL recognition. As many of these changes are located in the ß-pleated sheet floor of the groove, and presumably have no direct access to interaction with T cell receptors (TCR) (2), this has suggested the participation of endogenous peptides in allospecific T cell recognition. Characterization of new subtypes also provides information on the genetic mechanisms generating class I polymorphism and on the rules governing HLA diversification (1). In this report we summarize recent work from our lab, concerning the analysis of HLA-A2 antigens expressed on cells that are differentially recognized by allospecific CTL (3), and of HLA-A2 variants defined by isoelectric focusing (IEF) (4, 5).

PEPTIDE-MEDIATED CTL ALLORECOGNITION OF HLA-A2

In the first part of this work, we have analyzed the molecular basis of the reactivity of a HLA-A2 allospecific CTL line (CTL14) showing heterogeneity in the recognition of HLA-A2, through the characterization of differentially recognized HLA-A2 antigens. CTL14 was generated by J. van der Poel et al. (6) in the CML response of an HLA-A2⁻ individual (HLA-A1, 30, B8, 27) against PBL from the A2.4⁺ KNE individual (HLA-A1, 2, B8, 27), who expresses the $A*0207$ subtype. In addition to the stimulator cells, CTL14 recognized CLA cells ($A*0206$) and PBLs from 50% of the A2.1⁺ individuals analyzed in the study (Figure 1.A). These CTL did not recognize SCHU (A2.4⁺), KLO ($A*0208$), A2.2⁺, A2.3⁺ and a fraction of the A2.1⁺ cells.

First, we analyzed the HLA-A2 antigen expressed on SCHU cells, by comparative peptide mapping with the $A*0201$ antigen from JY. This technique is based on the metabolic labelling of cells expressing the known and the unknown A2 antigen with ¹⁴C and ³H labelled amino acids, respectively (7). Then the A2 proteins are isolated by immunoprecipitation and jointly digested with trypsin. The tryptic peptides are then fractionated by HPLC, allowing detection of the peptides carrying the amino acid differences between the HLA-A2 antigens being compared.

With this approach, we demonstrated that A2-SCHU differs from the A*0201 antigen

T Lymphocytes: Structure, Function, Choices, Edited by F. Celada
and B. Pernis, Plenum Press, New York, 1992

by an amino acid change of Phe-->Tyr at position 9 . This change is the only substitution that distinguishes A*0206 (from CLA line) from A*0201 (2,7). Therefore, we compared the A2-SCHU and A*0206 antigens. Figure 1.B shows that the Lys and Arg tryptic maps, which cover all the tryptic peptides except the carboxyl-terminal one, show no differences between both molecules, suggesting that they are biochemically identical. As this technique does not formally rule out the possibility of having highly conservative substitutions with no effect on the chromatographic profile of the corresponding peptides, we sequenced the cDNA coding for A2-SCHU. This was done by PCR amplification of full length cDNA transcripts, using 2 oligonucleotides mapping at the 3'-untranslated region and at the 5'leader exon of class I genes. The primers incorporate restriction sites for Eco RI and Hind III respectively, to allow cloning of the amplified material into adequate vectors (8). 5 of the pUC clones containing A2-SCHU cDNA were selected with an A2-specific probe and subcloned in M13 for sequencing. 5 M13 clones corresponding to the 5 different pUC clones were sequenced. The consensus sequence derived for A2-SCHU was identical to the cDNA sequence of A*0206 (2) (figure 1.C). Thus, the two differentially recognized cells, CLA and SCHU, express the same A*0206 antigen.

The structural identity between these two HLA-A2 antigens suggested that CTL14 could differentiate also A2.1⁺ cells on the basis of factors other than the primary structure of their HLA-A2 antigens. Thus we analyzed the A2.1 antigens from a recognized (ARC) and from an unrecognized cell (ESST) with the A2.1 antigen (A*0201) of JY, by

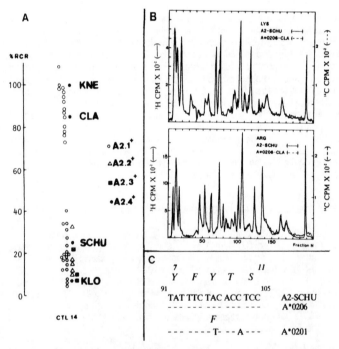

Figure 1. **A)** Recognition pattern of PBLs from HLA-A2⁺ individuals by the allogeneic line CTL14. Adapted from reference 6. **B)** Comparative peptide mapping of A2-SCHU with A*0206 from CLA. The Arg and Lys maps show the identity between both antigens. **C)** cDNA and deduced amino acid sequence of A2-SCHU in the region of difference with A*0201, compared with this antigen and A*0206 from CLA. All three sequences are identical outside this segment. (Reprinted by permission of Elsevier Science Publishing Co., Inc. from "Recognition of distinct epitopes on the HLA-A2 antigen by cytotoxic T lymphocytes," by J.J. Van der Poel, J. Pool, E. Goulmy, M.J. Giphart, J.J. van Rood, HUMAN IMMUNOLOGY, Vol. 16, pp. 247, Copyright 1986, by the American Society for Histocompatibility and Immunogenetics.)

comparative peptide mapping. The corresponding Lys and Arg maps showed no differences with the *A*0201* product, indicating that A2-ARC and A2-ESST were identical A2.1 antigens by this criterion (fig. 2).

These data imply that CTL14 can differentially recognize identical HLA-A2 antigens. The most likely explanation is that CTL14 recognizes an endogenous peptide presented by HLA-A2, that is only present in a fraction of the A2⁺ individuals. This could be due to a possible origin of the putative peptide from a polymorphic protein that is expressed in nearly 50% of the tested individuals. The possibility of polymorphism in the processing machinery generating different endogenous peptides in different individuals can not be ruled out. However, if this were so, one would expect to find this type of reactivity more frequently, which is not the case. Both explanations imply that some alloreactive CTL may recognize specific peptides presented by the MHC alloantigen, thus being allorestricted (9).

We have also analyzed three HLA-B7 variants defined by allospecific CTL clones (*B*0702*, B7.1 and B7.4) (10) and have shown that they were undistinguishable from the *B*0702* product (B7.2) by peptide mapping (A.R. Castaño, unpublished data). This suggests again that the CTL used to define these variants were allorestricted and could recognize endogenous peptides differentially expressed in the population, as presented by B7.2. The same situation has been found in the analysis of HLA class II antigens expressed in cells that were distinguished by DR allospecific CTL, in which no further polymorphism in the ß chain was found (11). Much indirect evidence suggests implication of peptides in allorecognition (12-20). However, its likely that a majority of the endogenous peptides are not polymorphic, so that alloreactive CTL recognizing these peptides would not reveal their allorestricted nature in the panel analyses used for their functional characterization. Finally, the large number of CTL precursors capable of developing into allorestricted CTL upon appropriate stimulation of unfractionated T cells with virus-infected allogeneic cells (21) further supports the quantitative importance of allorestriction in allospecific responses.

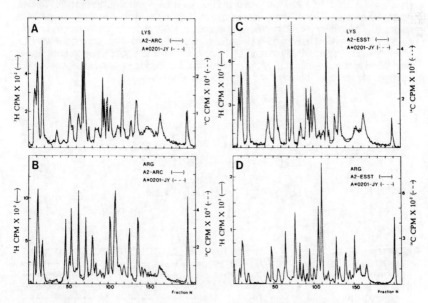

Figure 2. Comparative peptide mapping of A2-ARC and A2-ESST with the *A*0201* product. The Arg and Lys maps show structural identity with A2.1 in both cases. (Reprinted by permission of Elsevier Science Publishing Co., Inc. from "Recognition of distinct epitopes on the HLA-A2 antigen by cytotoxic T lymphocytes," by J.J. Van der Poel, J. Pool, E. Goulmy, M.J. Giphart, J.J. van Rood, HUMAN IMMUNOLOGY, Vol. l6, pp. 247, Copyright 1986, by the American Society for Histocompatibility and Immunogenetics.)

SELECTION-DRIVEN DIVERSIFICATION OF HLA-A2 SUBTYPES

Two previously unknown HLA-A2 variants were identified by IEF analysis in the Tenth Histocompatibility Workshop (22). These two new antigens, designated A*0204 (A2.4) and A*0211 (A2.5) are the most acidic HLA-A2 subtypes, differing from the A*0201 product in one and two charge units, respectively. Their structure was determined after PCR amplification, using the same strategy described for A2-SCHU. Total RNA from RML (HLA-A*0204, B51, DRw16, Dw22, DQw7) and KIME (HLA-A*0211, 32, Bw52, w61, Cw6) was used as starting material for sequencing each allele. The consensus sequences obtained from 4 A*0204 and 3 A*0211 M13-derived clones showed a single nucleotide change at position 362 (G-->T) in A*0204 and two nucleotide changes at position 290 (C-->T) and 292 (C-->G) in A*0211 relative to A*0201 (fig. 3.A). These base changes imply one amino acid replacement of Arg-->Met at position 97 in A2.4 and two amino acid replacements of Thr-->Ile at position 73 and His-->Asp at position 74, in A2.5. The nature of these changes accounts for the IEF pattern of these subtypes.

The only nucleotide substitution found in A*0204 suggests its origin by point mutation from A*0201. Other HLA antigens that have Met at position 97 have additional changes very close to this position (A68.1 or the Aw19 group) (23). Thus, an origin of A*0204 by gene conversion might be unlikely. A*0204 could represent an example of the contribution of recurrent point mutations to HLA diversification. As this subtype has been detected only in South American Indians but not in North American or Mexican Indians (22,24), it is possible that it originated locally among South American Amerindians.

The sequence of A*0211 is rather unusual among class I HLA antigens, as it has only been found in HLA-31 and Aw33 (25). The sequence identity in the region of difference from A*0201 between these 3 antigens (fig. 3.B) suggests that HLA-A31 or Aw33 may have acted as donor genes in a gene conversion event generating A*0211. The high proportion of Aw33 in South Asian populations, where A*0211 was first defined, favors this possibility (26). In addition, A*0211 has been recently found in an isolated Amerindian population in which HLA-A31 was one of the very few HLA-A antigens present (D. Watkins, personal communication). The presence of this allele in two distantly related populations raises the possibility of its origin as a result of independent, but analogous, gene conversion events. Thus, recurrent gene conversion might also contribute to HLA diversification. The possibility that A*0211 was imported into South America from Asiatic migrations cannot be ruled out at the present time, but it appears less likely because this subtype has not been detected in North American Indians.

Subtype	Nucleotide changes	Amino acid changes
A*0204	G_{362} ---> T_{362}	Arg_{97} --> Met_{97}
A*0211	C_{290} ---> T_{290}	Thr_{73} ---> Ile_{73}
	C_{292} ---> G_{292}	His_{74} ---> Asp_{74}

```
            70              77    281                    303
HLA-A2      H S Q T H R V D       CACTCACAGACTCACCGAGTGGAC
HLA-A*0211  - - - I D - - -       -----------T-G----------
HLA-Aw33    - - - I D - - -       -----------T-G----------
HLA-A31     - - - I D - - -       -----------T-G----------
A-CONSENSUS - - - - D - - -       ------------G-----------
```

Figure 3. **A)** Nucleotide and amino acid changes of A*0204 (A2.4) and A*0211 (A2.5) relative to A*0201. **B)** Comparison of the amino acid and nucleotide sequences of A*0211 with HLA-A alleles showing the same changes, in the region of difference with A*0201.

Residue 97, changed in A2.4, is located in the β-pleated sheet. Its side chain points towards the groove, taking part in the structure of pockets C and E (27). Residues 73 and 74, changed in A2.5, are located in the α1 α helix, pointing into the groove, and taking part on the formation of pocket C (27). The changes in both subtypes can affect peptide presentation by HLA-A2 and T cell recognition, therefore being functionally distinct alleles.

The spatial location of the polymorphic residues among HLA-A2 subtypes reveals a striking clustering of changes in a particular area of the β-pleated sheet floor. Of 13 polymorphic residues, 10 are found in the peptide binding groove (fig. 4). Four of them, residues 9, 95, 97 and 99, are located in two adjacent β strands at the α1-α2 interface, residue 9 being in contact with the three other residues. Among the 11 HLA-A2 subtypes there are 7 different amino acid combinations in these 4 residues. This equals the total number of combinations among all HLA-A antigens (fig. 5). As changes at all 4 positions can affect peptide presentation (2,28,29), their combination can modulate the fine specificity of peptide presentation by HLA-A2. Residues 73 and 74, changed in A2.5, are very close to this cluster, residue 74 being in contact with residues 95 and 97 (30). The His to Asp change originates a negatively charged pocket also contributed for by residues 9, 95 and 99 (31).

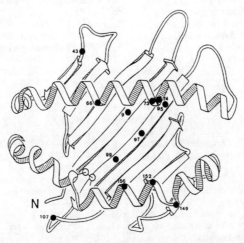

Figure 4. Spatial location of the polymorphic residues among HLA-A2 subtypes in the antigen binding site (27), except residue 236, in α3, which is changed in A*0209 (32).

SUBTYPE	POSITION			
	9	95	97	99
A*0201, 03, 09, 11	F	V	R	Y
A*0202	L	–	–	–
A*0205, 08	Y	L	–	–
A*0206	Y	–	–	–
A*0207	–	–	–	C
A*0210	–	–	–	F
A*0204	–	–	M	–

Figure 5. Patterns of polymorphism in ß-pleated sheet residues among HLA-A2 subtypes.

The observed clustering of polymorphic residues among HLA-A2 subtypes is different from that found in HLA-B27 subtypes. This concerns mainly the 74-81 segment of the $\alpha 1$ α helix, with the ß-pleated residues changed in HLA-A2 being conserved (1). This suggests that diversification of HLA antigens is driven by selection to alter specific areas within the peptide binding site. The driving force would be the effect that changes in these regions have on the particular peptide presenting specificity of a given HLA molecule. As different alloantigens bind distinct sets of peptides with specific structural motifs (33), selective pressures may favor variations at different locations, depending on the influence of these positions in the interaction with the particular set of peptides presented by the alloantigen. The interaction with antigenic peptides from environmental pathogens could dictate HLA subtype diversification in a selection-driven evolution. In this respect, the observed differences in ethnic distribution of HLA-A2 subtypes could be due to selection of those molecules that represent a selective advantage in dealing with local pathogens prevailing in each population.

ACKNOWLEDGEMENT

Supported by grant PB87/0347 from the Comisión Interministerial de Ciencia y Tecnología.

REFERENCES

1. López de Castro, J.A. "HLA-B27 and HLA-A2 subtypes: Structure, evolution and function". Immunology Today 10: 239 (1989).
2. Mattson, D.H., Shimojo, N., Cowan, E.P., Baskin, J.J., Turner, R.V., Coligam, B.D., Maloy, W.L., Biddison, W.E. "Differential effects of amino acid substitution in the ß-sheet floor and α2 helix of HLA-A2 on recognition by alloreactive viral peptide-specific cytotoxic lymphocytes". J. Immunol. 143: 1101 (1989).
3. Castaño, A. R., Lauzurica, P., Doménech, N, López de Castro, J.A. "Structural identity between HLA-A2 antigens differentailly recognized by alloreactive cytotoxic T lymphocytes". J. Immunol. 146: 2915 (1991).
4. Castaño, A. R., López de Castro. "Structure of the HLA-A*0204 antigen, found in South American Indians. Spatial clustering of HLA-A2 subtype polymorphism". Immunogenetics 34: 281 (1991).
5. Castaño, A.R., López de Castro. "Structure of the HLA-A*0211 (A2.5) subtype: further evidence for selection-driven diversification of HLA-A2 antigens". Immunogenetics "in press".
6. Van der Poel, J.J., Pool, J., Goulmy, E., Giphart, M.J. y van Rood, J.J. "Recognition of distinct epitopes on the HLA-A2 antigen by cytotoxic T lymphocytes". Human Immunol 16: 247 (1986).
7. Ezquerra, A, Doménech, N., van der Poel, J.J., Strominger, J.L., Vega, M.A. y López de Castro, J.A. " Molecular analysis of an HLA-A2 functional variant CLA defined by cytolytic T lymphocytes". J. Immunol. 137: 1642 (1986).
8. Watkins, D.L., Chen, Z.W., Hughes, A.L., Evans, M.G., Tedder, T.F. y Letvin, N.L. "Evolution of the MHC class I genes of the New World primate from ancestral homologues of human non-classical genes". Nature 346: 60 (1990).
9. Matzinger, P. y Bevan, M.J. "Why do so many lymphocytes respond to major histocompatibility antigens?". Cell. Immunol. 29: 1 (1977).
10. Van Seventer, G.A., Hiuis, B., Melief, C.J. y Ivanyi, P. "Fine specificity of human HLA-B7-specific cytotoxic T-lymphocyte clones. Identification of HLA-B7 subtypes and histotopes of the HLA-B7 cross-reacting group". Human Immunol. 16: 375 (1986).
11. Santamaría, P., Boyce, M.T., Lindstrom, A.L., Rich, S.S., Faras, A.J. y Barbosa, J.J. "Alloreactive T cells can distinguish between the same human class II MHC products on differente B cell lines". J. Immunol. 146:1822 (1991).
12. Marrack, P., and J. Kappler. "T cells can distinguish between allogeneic major histocompatibility complex products on different cell types". Nature 332:840 (1988).
13. Molina, I. J., and B. T. Huber. "The expression of a tissue specific self-peptide is required for allorecognition". J. Immunol. 144:2082 (1990).
14. Bernhard, E.J., Le, A.X., Barbosa, J.A., Lacy, E. y Engelhard, V.H. "Cytotoxic T lymphocytes from HLA-A2 transgenic mice specific for HLA-A2 expressed on human cells" J. Exp. Med. 168: 1147 (1988).

15. Epstein, H., Hardy, R., May, J.S., Johnson, M.H. y Holmes, N. "Expression and function of HLA-A2.1 in transgenic mice". Eur. J. Immunol. 19: 1575 (1989).
16. Shinohara, N, Bluestone, J.A., Sachs, D.H. "Cloned cytotoxic T lymphocytes that recognize an I-A region product in the context of a class I antigen". J. Exp. Med. 163:972 (1986).
17. Schendel, D. J. "On the peptide model of allorecognition: cytotoxic T lymphocytes recognize an alloantigen encoded by two HLA-linked genes". Hum. Immunol. 27:229 (1990).
18. Kievits, F. Ivanyi, P. "A subpopulation of mouse cytotoxic T lymphocytes recognizes allogeneic H-2 class I antigens in the context of other H-2 class I molecules". J. Exp. Med. 174:15 (1991)
19. Heath, W. R., M. F. Hurd, F. R. Carbone, and L. Sherman. "Peptide-dependent recognition of H-2K^b by alloreactive cytotoxic T lymphocytes". Nature 344:749 (1989).
20. Clayberger, C., Parham, P., Rothbard, J., Ludwig, D.S., Schoolnik, G.K. y Krensky. A.M. "HLA-A2 peptides can regulate cytolysis by human allogenic T lymphocytes". Nature 330: 763 (1987).
21. Kabelitz, D., W. R. Herzog, K. Heeg, H. Wagner, and J. Reimann. "Human cytotoxic T lymphocytes. III. Large numbers of peripheral blood T cells clonally develop into allorestricted anti-viral cytotoxic T cell populations in vitro". J. Mol. Cell. Immunol. 3:49 (1987).
22. Yang, S.Y. "Assignment of HLA-A and HLA-B antigens for the reference panel of B-lymphoblastoid cell lines determined by one-dimensional isoelectric focusing (ID-IEF) gel electrophoresis" En: Dupont, B. (ed) Immunobiology of HLA vol I. Immunogenetics and Histocompatibility. Springer-Verlag New York, pp 43 (1989)
23. Zemmour, J. and Parham, P. "HLA class I nucleotide sequences, 1991". Immunogenetics 33: 310 (1991).
24. Yang, S.Y. "Population analysis of class I HLA antigens by one-dimensional isoelectric focusing gel electrophoresis: workshop summary report". In: B. Dupont (ed.): Immunobiology of HLA. Vol. 1., pp. 309, Springer-Verlag, New York, 1989.
25. Kato, K., Trapani, J.A., Allopenna, J., Dupont, B., and Yang, S.Y.. "Molecular analysis of the serologically defined HLA-Aw19 antigens. A genetically distinct family of HLA-A antigens comprising A29, A31, A32 and Aw33, but probably not A30". J. Immunol. 143: 3371 (1989).
26. Imanishi, T., Akaza, T., Mura, T., Miyoshi, H., Kashiwase, K., Ina, Y., Tanaka, H., Tokunaga, K., Gojobori, T., and Juji, T. "Allele and haplotype frequencies of MHC loci estimate from panel data to the 11th International Histocompatibility Workshop: a preliminary analysis". In 11th International Histocompatibility Workshop. Data Analysis Book II p. 433. (1991)
27. Saper, M.A., Bjorkman, P.J. y Wiley, D.C. "Refined structure of the human histocompatibility antigen HLA-A2 at 2.6 A resolution". J. Mol. Biol. 219: 277 (1991).
28. McMichael, A.J., Gotch, F.M., Santos-Aguado, J., Strominger, J.L. "Effect of mutations and variations of HLA-A2 on recognition of a virus peptide epitope by cytotoxic T lymphocytes". Proc. Natl. Acad. Aci. USA 85: 9194 (1988).
29. Robbins, P.A., Lettice, L.A., Rota, P., Santos-Aguado, J., Rothbard, J., McMichael, A.J., Strominger, J.L. "Comparison between two peptide epitopes presented to cytotoxic T lymphocytes by HLA-A2. Evidence for discrete locations within HLA-A2". J. Immunol. 143: 4098 (1989).
30. Bjorkman, P.J., Saper, M.A., Samraoui, B., Bennett, W.S., Strominger, J.L., and Wiley, D.C. "Structure of the human class I histocompatibility antigen, HLA-A2". Nature 329: 506-512 (1987).
31. Garrett, T.P.J, Saper, M.A., Bjorkman, P.J., Strominger, J.L. y Wiley, D.C. "Specificity pockets for the side chains of peptide antigens in HLA-Aw68". Nature 342: 692 (1989).
32. Castaño, A.R., Ezquerra, A., Doménech, N., López de Castro, J.A. "An HLA-A2 population variant with structural polymorphism in α3 region". Immunogenetics 27: 143 (1988).
33. Falk, K., Rötzscheke, O., Stevanovic, S., Jung, G. y Rammensee, H.G. "Allele-specific motifs revealed by sequencing of self-peptides eluted from MHC molecules". Nature 351: 290 (1991).

10

ON THE ANTIGENICITY OF ANTIBODY IDIOTYPES

Kristian Hannestad

Department of Immunology, Institute of
Medical Biology, University of Tromsø
School of Medicine, Tromsø

SUMMARY

1. The T helper (Th) cell responses against isologous myeloma protein V-315 domains were regulated by MHC-linked immune response (IR) genes indicating that Id are recognized like ordinary protein antigens. 2. Immunization with synthetic idiopeptides revealed that the Th focused on three point mutated amino acids in CDR3 of the L-chain of M315 indicating that autoimmunogenicity may be acquired by mutations. 3. Igh-1b specific $CD4^+8^-$ T cell clones of the Th1 subset selectively suppressed the cognate Igh-1b allotype in vivo indicating that T cells can interact with B cells presenting peptides from endogenous Ig. 4. The germ-line gene encoded complete myeloma protein J558 was non-immunogenic but its free L-chain was immunogenic for Th, indicating that an idiopeptide that is cryptic in the complete Ab may become available when the L-chain is free. 5. IgM mAb formed spontaneous complexes with albumin in bovine serum-containing medium, and albumin in the complex dramatically augmented immunogenicity of IgM. Therefore hybridomas producing mAb to be used for therapy should be grown in serumfree medium to prolong the half-life of the Ab. 6. One native IgM mAb, free of adjuvant and foreign serum protein, has been found that elicited humoral anti-Id responses when given as a single s.c. or i.p. injection of 100 μg. This indicated that Id-determinants alone can be immunogenic under conditions that are closer to physiological than adjuvant-dependent responses. 7. Possible roles for Id of IgM mAb as cardinal autoantigens are outlined.

INTRODUCTION

Since the diversity of Ab V-domains is extremely high[1], the major part of the Ab repertoire may not reach toleragenic concentrations. If so, many clonotypic Ab are "non-self" like ordinary foreign proteins. To what extent can an immune system recognize and respond to its own soluble native V-domains (idiotypes -Id)? The answer to this question is unknown and has implications for the regulation of Ig production by normal B cells[2], growth inhibition of B cell neoplasias[3-5], practical application of therapy with human or "humanized" mAb[6], and the role of Id as autoantigens in autoimmune disease (see Discussion).

The present communication reviews several studies from this laboratory aimed at gaining information about recognition of isologous Ig by Th helper cells (Th) and the presentation of endogenous Ig by B cells to Th.

T Lymphocytes: Structure, Function, Choices, Edited by F. Celada
and B. Pernis, Plenum Press, New York, 1992

Cognate Th-B interactions reveal activity of Th

Studies from the laboratory of H.N. Eisen revealed that several mouse IgA myeloma proteins including one called 315, elicited a humoral anti-Id response in syngeneic mice provided they were injected repeatedly in large amounts and together with a powerful adjuvant[7,8]. Work from our laboratory has demonstrated that myeloma protein 315 (isotype IgAλ2, anti-TNP-lysine) also elicits a T helper (Th) cell response[9,10]. The Id-primed Th can be detected by their ability to interact with hapten (NIP) primed B cells.

To detect Th-B interactions we exploited the "three mice" experimental system of Mitchison[11]: <u>mouse A</u> is primed i.p. with NIP-ovalbumin (OA) in CFA; <u>mouse B</u> is primed with the antigen recognized by Th (called the carrier). In our case the carrier was either the variable (V) domain of the light chain (V_L) or of the heavy chain (V_H) of myeloma protein 315. NIP-primed B cells from mouse A and carrier (Id)-primed popliteal lymphnode cells from mouse B were transferred to <u>mouse C</u> (given 500 rad beforehand) where they may interact following a boost of mouse C with NIP-V_L or NIP-V_H. By changing the carrier of the boost antigen, the OA primed splenic Th of mouse A are excluded from the Th-B interactions in mouse C whereas Id-specific Th are offered an opportunity to help NIP-specific secondary B cells that have captured a NIP-conjugated V-domain and present idiotypic peptides from V_H or V_L. As a result of this linked (cognate) interaction the anti-NIP B cells can mature to high rate Ab producing plasma cells. Thus, when mouse C is bled 9 days after the boost and the serum is analyzed for anti-NIP activity, a high anti-NIP response signifies that the Th have engaged in cognate interactions with idiopeptides presented by the NIP-specific B cells. No or a minimal anti-NIP response indicates that no such interactions have occurred.

This system has several advantages: a) it is very sensitive, b) it is in vivo based, c) it depends on a biologically important T cell effector activity, i.e. help for B cells, d) it reflects the function of a polyclonal secondary Th population, and e) by using isolated V-domains for priming and paired (V_H+V_L) domains, i.e. Fab fragments, for the boost we made sure that the Th-B interactions reflected idiopeptides that were generated from both free and assembled V domains. This was important since subsequent studies of another myeloma protein (J558) revealed that in this case the Th only responded to free L-chains, not to the complete syngeneic J558 antibody (see below).

The responses against the V-domains of M315 are regulated by H-2 linked immune response (IR) genes

Using the experimental approach described above we obtained data that are summarized in Table 1. From these results it may be concluded that:

1. H-2^d and H-2^s confer high responsiveness to V_L^{315} (lines 1a,5,11). For H-2^d the high responder gene maps to E (lines 1a,1b) whereas for H-2^s it maps to the A subregion (line 11).
2. Non-H-2 background gene(s) of B10 strain quench the Th response to V_L (line 9). The explanation is not clear.
3. H-2^k confers high responsiveness to V_H^{315} (lines 2,4,8,10) and the gene maps to the A-subregion (lines 7,8).
4. H-2^b confers low responsiveness to both V_H and V_L (lines 3,6).

Thus, myeloma protein 315 appears to be recognized like a conventional protein antigen by Th, i.e. the Th response required presentation by MHC molecules (reviewed in ref. 12).

Table 1. T helper cells primed with V_H^{315} or V_L^{315} and boosted with NIP-Fab^{315} augment secondary adoptive anti-NIP responses as a function of H-2 recombinant haplotypes.

Strain		I-A	I-E	Ag used for Th priming	
				V_H	V_L
1a.	BALB/c	d	d	low[a]	high
1b.	D2.GD	d	-[b]	ND	low
2.	BALB.H-2^k	k	k	high	low
3.	BALB.H-2^b	b	-	low	low
4.	C3H	k	k	high	low
5.	C3H.H-2°	d	d	low	high
6.	C3H.SW	b	-	low	low
7.	B10	b	-	low	low
8.	B10.A(4R)	k	-	high	low
9.	B10.D2	d	d	low	low
10.	A.TL	k	k	high	low
11.	A.TH	s	-	low	high

[a]Humoral anti-NIP response in adoptive host.
[b]I-E not expressed. The data are summarized from ref. 9 and 10.

The Th response against the L-chain of myeloma protein 315 (L^{315}) is focused on a hypermutated segment in CDR3

L^{315} contains a cluster of three point mutations in CDR3 of the $\lambda 2$ gene-segment: $Y^{94}S^{95}T^{96} \rightarrow F^{94}R^{95}N^{96}$ (Ref. 13). (The first amino acid of the J-region is position 98.) To explore the significance of these point mutations for Th responses to a challenge with NIP-L^{315}, we then synthesized three peptide analogues, each containing the germ-line gene (non-mutated) amino acid in one of the three mutated positions. The animals were primed with the peptide analogue, and their lymphnode cells were transferred together with NIP-primed B cells to irradiated recipients which were boosted with NIP-L^{315}. The results are shown in Table 2.

Th primed with synthetic peptide analogues where R^{95} and N^{96} had been exchanged with S^{95} and T^{96}, did not cross-react with L^{315}. Th of four out

Table 2. Point mutated amino acids in CDR3 are critical for Th recognition of the $\lambda 2^{315}$ L-chain.

Position 94-96 sequence of		Recipient
Th-priming	boost antigen	anti-NIP
peptide 91-108	(L^{315})[a]	response
FRN	FRN	high
YRN	FRN	high[b]
FSN	FRN	none
FRT	FRN	none

[a]The antigen was NIP-conjugated
[b]Two of six animals did not respond.
The data are summarized from ref. 14.

of six animals tolerated the conservative $F^{94} \rightarrow Y^{94}$ substitution, but two animals did not cross-react. Thus, the mutated amino acids R^{95} an N^{96}, and to a lesser extent F^{94}, were critical for an idiotypic site of L^{315} that was recognized by Th. This mutated CDR3 site is the only one that Th can recognize because we observed no responses against the L^{315} cleavage peptide 1-87 (Ref. 15). The specificity for mutated amino acids of this polyclonal Th populations as well as their MHC-restriction was subsequently confirmed by the proliferative responses of cloned T cell lines[16,17].

Igh-1b-specific CD4+8− T cell clones of the Th1 subset selectively suppress the Igh-1b allotype in vivo

The studies described above demonstrated that B cells can present idiopeptides from soluble syngeneic Ig to Th. This raised the question of whether a B cell processes and presents the Ig it synthesizes, i.e. its endogenous Ig, to MHC class II restricted T cells. An obstacle to studies of interactions between normal unmanipulated B cells presenting idiopeptides and Id-specific T cells is the low prevalence of B cells expressing a given Id; the target clonal B cell and its idiotypic product are usually very hard to detect. In contrast, B cells expressing a given Ig C-region allotype are more abundant. We therefore established a BALB/c (Ig^a,$H-2^d$) T cell line and clones specific for the Igh-1b allotype. The cells were CD4+8− of the Th1 subset ($I1-2^+$,$IL-4^-$,$INF\gamma^+$,TNF^+), and they recognized Igh-1 (IgG2a) of the b allotype (Igh-1b) together with I-A^d.

The Ig allotype specific T cells specifically suppressed B cells synthesizing Igh-1b in vivo. Thus, in 12 out of 15 six-week-old (BALB/c x B10.D2)F1 mice injected neonatally with Igh-1b-specific T cells, the serum Igh-1b-concentrations were <5% of the levels in controls[2]. Thus, allotype suppression can be accomplished solely by adoptive transfer of Igh-1b-specific CD4+ T cells. The in vivo suppression was specific for Igh-1b+ B cells because the recipients' levels of Igh-1a and Igh-4b (IgG1^b) were unaffected. The results indicated that suppression resulted from cognate interactions between allopeptide specific TcRαß+ T cells and normal unmanipulated B lymphocytes presenting their endogenous Igh-1b in association with MHC class II. The data support the possibility that normal B cells presenting endogenous idiopeptides can be suppressed by idiopeptide-specific T cells of the Th1 subset in vivo. Id-specific Th1 cells may represent a defense mechanism against B cell neoplasms by a similar mechanism.

A germ-line gene encoded Ab (J558) is non-immunogenic but its free λ1 L-chain is immunogenic for Th

The BALB/c myeloma protein J558 (IgA λ1, anti-Dextran) does not contain somatic point mutations, i.e. it is germ-line gene encoded. Unlike myeloma protein 315 the complete J558 antibody did not elicit Th responses, even when emulsified in CFA (Table 3, line 4,5). In marked contrast, its free λ1 L-chain was immunogenic in BALB/c mice (Table 3, line 1), but the primed Th did not cross-react with λ1-chains paired with V_H^{J558} (line 2).

Antibodies identical to J558 are frequently elicited in response to Dextran[19]. Thus, the simplest explanation for these results is that J558 is abundant in BALB/c mice due to immunization with dextran from bacteria in the gut. Therefore, the animals are tolerant to the complete J558. The free L^{J558}-chain, by contrast, is present in subtoleragenic concentrations. We hypothesize that when the free L^{J558} chain is processed by APC, one or more unique idiopeptides are generated that differ from L^{J558} peptides originating from the complete J558. The immune system is not

Table 3. Th recognize the free but not the assembled
L-chain of myeloma protein J558.

Th-priming antigen	Boost antigen[a]	Recipient anti-NIP response
1. Free L^{J558}	Free L^{J558}	high
2. "	Fab^{J558}	none
3. "	Free L^{315}	none
4. $Complete^{J558}$	Fab^{J558}	none
5. "	Free L^{J558}	none

[a]The antigen used for boosting was NIP-conjugated.
The data are summarized from ref. 18.

tolerant to these unique peptides and Th can therefore respond to the
free L^{J558}.

If this is correct the J558 system provides an example of how a
cryptic peptide can become available for T cells, i.e. by dissociation of
an antigen into its subunits. Conceivably such cryptic peptides may be
generated naturally in B cells by processing of endogenous H and L chains
before they have assembled into complete Ig molecules intracellularly. In
this case cryptic idiopeptides may be presented by MHC on, for example,
neoplastic B cells and serve as tumor specific transplantation antigens.
In this setting, therefore, cryptic idiopeptides may assume biological
function.

Does an immune system respond to adjuvant-free Id from the pre-immune repertoire?

If the Ids of B cell clones engaged in an immune response frequently
stimulate anti-Id T- and B-cells, it should be possible to show similar
anti-Id responses by injecting adjuvant-free mAb. It is unclear, however,
whether isologous mAb are commonly or only rarely immunogenic in the
absence of adjuvants. We have previously demonstrated that the native 7S
monomer, but not the polymer, of IgAλ2 myeloma protein 315 purified from
ascites stimulated CD4[+] Th-dependent anti-Id Ab[20]. Since this mAb has
many point mutations it may be a special case, and it is of interest to
examine to what extent an immune system can respond to V-domains from the
primary (pre-immune) repertoire. Studies in the 1970s indicated that many
isologous IgA mAb are immunogenic provided they were injected repeatedly
in large amounts and together with powerful adjuvants[7,8]. The Th cell
response against an idiopeptide of one of these (M315) has been described
above.

However, for many IgG mAb CFA was not sufficient and immunogenicity
was only achieved after they had been polymerized with glutaraldehyde[21,22], linked to a foreign protein[23] or hapten[24,25] or adsorbed to
bacteria[26]. These maneuvers of artificial enhancement of immunogenicity
denatured the Ig and/or introduced new (nonself) determinants for T
cells. The requirement for Th recognition of native idiopeptides was
therefore short-circuited. Consequently, the key question still remains
as to what extent syngeneic native Ab free of adjuvant and foreign
proteins are immunogenic under more physiological conditions.

To study this question we made 21 BALB/c hybridoma anti-TNP mAb of
IgMλ2 isotype. Unexpectedly, we found that all mAb from supernatants
containing fetal bovine serum (FBS), and purified by their affinity for

DNP-lysine-sepharose, had formed spontaneous complexes with BSA. Several lines of evidence supported this conclusion. First, a two-site sandwich ELISA demonstrated BSA-IgM heterocrosslinks between plate-bound anti-BSA and soluble anti-IgM Ab; in this assay IgM should have been washed away if it were not bound to BSA. Second, mice immunized with IgM purified from 10% FBS supernatants made anti-BSA Ab. Finally, strong humoral anti-Id responses followed immunization with 10 μg adjuvant-free IgM[6] isolated from 10% FBS-supernatants. By contrast, 100 μg of IgM[6] isolated from ascites or serumfree medium was not detectably immunogenic; even a 50 μg boost failed.

The most likely explanation for the striking difference in immunogenicity of IgM[6] from the various sources is that since BSA (the "carrier") was linked to IgM[6] it was interiorized together with IgM[6] in B cells bearing anti-Id receptors. As a result the anti-Id B cells came under the powerful influence of Th cells specific for BSA peptides together with MHC class II on these B cells, leading to a strong augmentation of their anti-Id response, i.e. a carrier effect[11]. As the IgM anti-DNP mAb did not cross-react with platebound BSA in ELISA, the complexes with BSA probably developed by other means than an Ab-Ag reaction. Previous reports[27,28] have documented complexes between human albumin and human IgM and presented evidence suggesting that disulfide interchange reactions were responsible.

Twelve of the 21 mAb isolated from 10% FBS supernatants elicited IgG1 anti-Id Ab following a single s.c. injection of only 10 μg adjuvant-free soluble IgM. The presence of BSA-IgM complexes probably explains why so many IgM mAb were immunogenic. Three of the twelve hybridoma IgM were therefore adapted to grow in serumfree medium and in ascitic form and then purified.

Interestingly, a 100 μg s.c. dose of adjuvant-free IgM[3] isolated from ascites and from serumfree medium did induce IgG1 anti-Id Ab in three out of five mice. A 50 μg s.c. boost led to a strong secondary response in all five animals, ranging from 0.8-350 μg ml^{-1} anti-Id Ab. Thus, the responses against IgM[3] were substantial and the hapten DNP-glycine specifically blocked the Id-anti Id reaction. These results indicate that there exists some isologous IgM mAb, such as IgM[3], whose Id determinants alone can elicit strong anti-Id responses even in the absence of adjuvant[29].

It is likely that these Id determinants are located in CDR3 of the H chain due to the enormous diversity of this region. However, we do not yet have direct evidence for this. It is hard to rule out the theoretical possibility that IgM[3], after it was injected into BALB/c mice, combined specifically in vivo with a foreign (environmental) antigen which provided carrier determinants. This is unlikely because the mice were healthy and no evidence was found that IgM[3] was polyspecific. The immunogenicity of a single small dose of adjuvantfree IgM-associated Id supports the idea that similar responses can occur physiologically.

DISCUSSION

A major conclusion from the studies presented here is that certain isologous monoclonal Ig can be recognized like ordinary foreign protein antigens. Several factors are likely to limit this type of anti-Id responses (Table 4).

Does there exist an Id-network based on Th recognition of idiopeptides? The observations with IgM[3] support this idea, but it remains to be proven. If it exists it is likely to be incomplete, i.e.

Table 4. Factors that limit Th dependent anti-Id responses.

---

1. Natural abundance of some clonotypic Ab induces non-responsiveness (Myeloma proteins TEPC15 and J558).
2. MHC-linked IR gene defect (myeloma protein 315).
3. The amount of Id produced naturally may be too small to elicit an immune response.
4. The amount of immunogenic idiopeptide generated by APC may be suboptimal (cryptic idiopeptides).
5. Gaps in the T cell repertoire may exist for some idiopeptides.

---

certain B cell clones, like IgM[3], could come under the powerful influence of idiopeptide-specific Th cells and anti-Id Ab, whereas the Id of other clones would not be recognized. In other words, the expansion of an Id is not necessarily linked with the generation of anti-Id. Clones that proliferate extensively and secrete substantial amounts of Id that is efficiently processed to yield immunogenic idiopeptides should be particularly prone to become targets for Th dependent anti Id-responses. It is not clear, however, whether in this setting Id production would be stimulated or suppressed. This could depend on the lymphokines secreted by Th engaged in such interactions or their cytolytic capacity (see ref. 2).

The observations concerning IgM[3] imply that Id of certain IgM may play an important role in autoimmunity. Consider for example a clone secreting an IgM Ab that bears serologically defined Id which mimics a conventional non-Ig autoantigen, i.e. it is an internal image of the autoantigen[30]. Why should the secreted Id (the internal image) be more potent than the autoantigen with respect to stimulating an autoAb-response? The answer could be that if the autoantigen-mimicking IgM mAb possesses idiopeptides capable of stimulating Th, it will be recognized like an ordinary protein antigen by collaboration between anti-Id B cells and idiopeptide + MHC class II specific Th. The outcome of such linked recognition could be an anti-Id Ab response that cross-reacts with the autoantigen. By contrast, non-Ig autoantigens lack the extensive diversity of CDR3; they have therefore usually tolerized Th and fail to elicit autoAb responses.

Second, Id may provoke disease by bystander injury. Thus, B cells bearing certain autoimmunogenic V-regions may infiltrate a target organ, for example due to (transient) microbial invasion and local production of cytotokines chemotactic for lymphocytes. Local deposition of autoimmuno-genic IgM (Id) follows, with priming of Id-specific B- and T cells, Id-antiId interactions, local inflammation and injury. Thus, the Ab (Id) may trigger inflammation not by its Ab function but by the immune response directed against its V-regions. According to this hypothesis, an activated and proliferating B cell clone could challenge the immune system like an infectious agent.

Since we only examined three IgMλ mAb that were free of foreign protein antigen, the frequency of clones with strongly autoimmunogenic Id from the pre-immune repertoire is unknown. But even if it is low, say one in a hundred, it could still be highly significant for autoimmunity in view of the large size of this repertoire.

REFERENCES

1. S. Tonegawa, Nature 302:575 (1983).
2. K. Bartnes and K. Hannestad, Eur. J. Immunol. 21:2365 (1991).

3. R.G. Lynch, R.J. Graff, S. Sirisinha, E. Simms, and H.N. Eisen, Proc. Natl. Acad. Sci. USA 69:1540 (1972).
4. S.L. Brown, R.A. Miller, S.J. Horning, D. Czerwinski, S.M. Hart, R. McElderry, T. Basham, R.A. Warnke, T.C. Merigan, and R. Levy, Blood 73:651 (1989).
5. C. Chan, F. Soehnlen, and G.P. Schechter, Blood 76:1601 (1990).
6. R.D. Mayforth and J. Quintans, N. Engl. J. Med. 323:173 (1990).
7. S. Sirisinha and H.N. Eisen, Proc. Natl. Acad. Sci. USA 68:3130 (1971).
8. N. Sakato and H.N. Eisen, J. exp. Med. 141:1411 (1975).
9. T. Jørgensen and K. Hannestad, J. exp. Med. 155:1587 (1982).
10. T. Jørgensen, B. Bogen, and K. Hannestad, Scand. J. Immunol. 21:183 (1985).
11. N.A. Mitchison, Eur. J. Immunol. 1:18 (1971).
12. R.H. Schwartz, Advances Immunol. 38:31 (1986).
13. A. Bothwell, M. Paskind, M. Reth, T. Imanishi-Kari, K. Rajewsky, and D. Baltimore, Nature 298:380 (1982).
14. K. Hannestad, G. Kristoffersen, and J.P. Briand, Eur. J. Imm. 16:889 (1986).
15. T. Jørgensen, B. Bogen, and K. Hannestad, J. exp. Med. 158:2183 (1983).
16. B. Bogen, B. Malissen, and W. Haas, Eur. J. Immunol. 16:1373 (1986).
17. B. Bogen, R. Snodgrass, J.P. Briand, and K. Hannestad, Eur. J. Immunol. 16:1379 (1986).
18. B. Bogen, T. Jørgensen, and K. Hannestad, Eur. J. Imm. 16:353 (1986).
19. B. Newman, J. Liao, F. Gruezo, S. Sugil, E. Kabat, M. Tarii, B. Clevinger, J. Davie, J. Schilling, M. Bond, and L. Hood, Mol. Immunol. 23:413 (1986).
20. G. Kristoffersen and K. Hannestad, Eur. J. Immunol. 18:1785 (1988).
21. V. Yakulis, N. Bhoopalam, and P. Heller, J. Immunol. 108:1119 (1972).
22. L.S. Rodkey, J. exp. Med. 139:712 (1974).
23. P.S. Fraker, L. Cicuriel, and A. Nisonoff, J. Immunol. 113:791 (1974).
24. G.M. Iverson, Nature 227:273 (1970).
25. E.A. Janeway and W.E. Paul, Eur. J. Immunol. 3:340 (1973).
26. K. Eichmann, Eur. J. Immunol. 2:301 (1972).
27. J.F. Heremans and M.-Th. Heremans, Acta Med. Scand. 170. Suppl. 367:27 (1961).
28. M. Mannik, J. Immunol. 99:899 (1967).
29. K. Hannestad, K. Andreassen, and G. Kristoffersen, Eur. J. Immunol. (in press).
30. N.K. Jerne, Ann Immunol. (Paris) 125C:373 (1974).

11

IDIOTYPE INTERACTIONS BETWEEN T CELLS

Benvenuto Pernis

Department of Microbiology
College of Physicians and Surgeons
Columbia University
New York, NY 10032

INTRODUCTION

Interactions between T cells mediated by their clonotypic receptors (idiotype interactions) have long been postulated as an extension of network concepts (1,2). Moreover, the possibility that these interactions may mediate different aspects of the complex regulation of T cells, as part of an integrated system, has been considered (3,4).

A recent review (5) that summarizes the functional implications of idiotype interactions between T cells is a clear indication of the importance, but also of the complexity, of this aspect of immunology. Research in this field has been hampered by the lack of knowledge, and sometimes lack of consideration, of the cellular and molecular events that must occur to allow mutual recognition among T cells, based on clonotypic recognition of the T cell receptor (TCR). This paper will deal with some facts that support the existence of regulatory idiotypic interactions between T cells *in vivo*, and will then include an analysis of what may be the cellular and molecular events that may allow the processing and presentation of the TCR by one T cell so that it may be specifically recognized by the TCR of another T cell.

EVIDENCE FOR IMMUNOREGULATION MEDIATED BY CLONOTYPIC (TCR) MOLECULES

It has long been postulated that specific interactions between T cell clones are fundamental in the regulation of the function of the immune system. Recent experimental work on the stimulation of specific suppressive T cells by other T cells activated with various antigens, as well as recent advances in the knowledge of antigen processing and presentation, allow the formulation of the hypothesis (6-8) that, under certain conditions, activated T cells

T Lymphocytes: Structure, Function, Choices, Edited by F. Celada
and B. Pernis, Plenum Press, New York, 1992

process their own T cell receptors (TCR) and present their idiotypic sequences to other complementary T cells that exert a regulatory function.

From the vast literature in this field we shall only quote recent contributions that are more directly relevant to the hypothesis outlined above.

Experimental allergic encephalomyelitis (EAE) can be induced, in different species, by immunization with a myelin basic protein (MBP). It has been shown that the disease is due to a reaction of CD4$^+$ T cells that recognize a limited set of peptides from the MBP, presented by class II MHC molecules. A remarkable finding borne out by the study of these MBP-recognizing T cells is that the diversity range of their TCR is very limited, both in the rat [Happ and Heber-Katz (9)] and in the mouse [Acha-Orbea et al. (10)]. The reasons for this restricted heterogeneity of the T cell clones that are responsible for the experimental disease are not clear; however, this restriction allows us to understand how an equally restricted reaction of other T cells, that recognize the idiotype of the cell receptors of the encephalitogenic clones, can actually be induced by immunization with cells stimulated by MBP.

The first observations in this direction were those of Ben-Nun et al. (11), who have seen the syngeneic rats inoculated with a highly encephalitogenic T cell line, propagated in vitro by continuous immunization with MBP, become resistant to the pathogenic effects of MBP, provided that the pathogenic capacity of the cells themselves was eliminated with irradiation or with mitomycin C. It is notable that the "vaccine" cells were boosted in vitro with antigen for 72 hours immediately before the irradiation (or mitomycin C treatment) and the intravenous injection in normal animals. Ben-Nun et al. explained their findings through an anti-idiotypic interaction between T cells.

Subsequently the same group confirmed and extended their findings, emphasizing that the T cells used for vaccination had to be activated, either by antigen or by mitogen, and that best results were obtained if the cells were fixed with glutaraldehyde or treated with high hydrostatic pressure (12). On the same direction, Sun et al. (13) have isolated CD8$^+$ T cell lines that specifically lysed live S1 cells (an established encephalitogenic CD4$^+$ line). These cytolytic cells expressed full length mRNA for TCR alpha and beta chains; surprisingly, their activation and their cytolytic activity was not blocked by antibodies directed against class I MHC nor by anti-CD8 antibodies.

More direct evidence for a role of TCR-related peptides in the regulation of EAE has been provided by Vanderbark et al. (14) and by Howell et al. (15). Both groups showed that immunization with (different) synthetic peptides derived from the variable region of the TCR of encephalitogenic cells provided protection in rats against experimentally induced EAE. An effect of TCR peptides on ongoing EAE has also been reported (16).

CD8$^+$ T CELLS IN THE REGULATION OF AUTOIMMUNE DISEASES

The data in the previous section indicate that the CD8$^+$ T cells may exert a regulatory role in autoimmunity. Recent work in our laboratory (17) supports this possibility. We have seen that depletion of CD8$^+$ T cells in vivo, performed in mice with administration of anti-CD8 antibodies, eliminates the resistance to a second episode of allergic encephalomyelitis that is otherwise dominant in mice that have overcome a first acute episode of the disease. In particular we have induced EAE in B10.PL mice by immunization with the acetylated 1-9 amino terminal peptide of MBP emulsified in complete Freund's adjuvant and accompanied by pertussis protein. These animals, after having spontaneously overcome an acute episode of encephalomyelitis, are almost completely resistant to the induction of a second episode induced in the same way. Depletion of CD8[+] T cells eliminates this resistance. On the other hand, depletion of CD8[+] T cells before the first episode does not affect the course of the disease. We interpret these observations in the sense that the encephalitogenic T cells responsible for the acute episode induced a satellite response of anti-idiotypic CD8[+] T cells capable of preventing a second episode of encephalomyelitis.

Similar observations on the role of CD8[+] T cells in the immunoregulation of EAE were performed by Koh et al. (18) in CD8 knock-out mice.

It is noteworthy, in addition, that spontaneous autoimmune diabetes has been observed by Faustman et al. (19) in mice in which the gene for beta-2 microglobulin had been knocked out. In these mice there are no class I MHC molecules and, consequently, no CD8[+] T cells; so the observations of Faustman et al. indicate a role of class I MHC or CD8[+] T cells, or both, in the prevention of autoimmunity.

The simplest, but by no means the only, interpretation of the data reported in this section is that activated T cells can process and present their own T cell receptor (TCR) as peptides complexed with their membrane class I MHC. These complexes could then be appropriate and specific targets for the TCR of regulatory (suppressor) CD8[+] T cells.

Since murine lymphocytes do not express class II MHC molecules, the only presenting molecules for TCR peptides would be class I: consequently the only other T lymphocyte that could recognize them would be the CD8[+] T cells, since the CD8 co-receptor is the one that focuses the TCR on antigens (peptides) presented by class I MHC. This would be a simple explanation of why the anti-idiotype category of suppressor cells is, at least in the mouse, composed of CD8[+] T cells. In other species in which T lymphocytes also can express class II MHC, CD4[+] anti-TCR idiotype suppressors might also be found.

THE PROBLEMS OF THE PRESENTATION OF THE T CELL RECEPTOR BY ACTIVATED T CELLS

Specific (clonotypic) interactions between T cells might, at first sight (5) appear as network interactions, in principle similar to those that can occur between antibodies. However, they are quite different due to the limitations imposed on antigen recognition by the TCR, that require its presentation, in principle as a peptide, bound to a defined cleft on the presenting (histocompatibility) molecules. These limitations imply unidirectionality in TCR-mediated interactions between T cells (one cell recognizes and the other is being recognized) and require or allow different cellular events and/or membrane accessory molecules to play a role in this recognition process, eventually turning it into a part of an elaborate and appropriately tuned immunoregulatory system.

The first step that we must take to study the clonotypic (idiotypic) interactions between T cells and their role in immunoregulation, is to consider the problem of the processing and presentation of the T cell receptor (20).

It is likely that self proteins, as many viral proteins, are fragmented in the cytosol and the derived peptides are transferred across the membranes of the endoplasmic reticulum (ER), after which the peptides may bind to the nascent class I MHC molecules. However, we have no clues to evaluate this possibility for the TCR α and β chains.

Actually, since both chains have signal sequences and are co-translationally transferred to the lumen of the ER, it is unlikely that the cytosol is a major site of TCR chain processing.

On the other hand, it has been established that α and β chains, which are synthesized in excess of the CD3 complex chains γ, δ, ϵ and ζ (zeta), are degraded in the cytoplasm of T cells, so that up to 85% of the newly synthesized TCR α and β chains never reach the cell surface (21). The degradation of this excess of α and β TCR chains takes place in the ER (22,23); molecules that escape this degradation and still are not part of the complete multichain receptor ($\alpha\beta$-$\gamma\delta\epsilon\zeta^2$) are diverted, at a trans-golgi station, to lysosomes where they are finally degraded (21). The degradation in the ER can produce peptides which, if they are of the appropriate size and sequence, may bind to the nascent class I MHC molecules that are being assembled in the ER itself and from there be transported to the cell membrane as a complex with class I MHC, in principle capable of being recognized by CD8+ cells.

It has also been established (24,25) that activated T cells show a prominent phenomenon that is not observed in any other kind of cells; namely that they actively endocytose and recycle their membrane class I MHC molecules through acidic (pH 5.5) endosomes. The fact that the phenomenon is peculiar to activated T cells suggests that its function is part of the physiology of these cells and of these cells only. We are now working on the hypothesis that it is involved in the presentation of the T cell receptor peptides on class I MHC. It has been recently suggested (26) that the association of antigenic peptides with class I MHC may occur "in a more distal exocytic compartment" than the ER. A simple possibility would be that TCR peptides generated in the ER or beyond are transported through the exocytic pathway in to the recycling endosomes where they could form complexes with the class I MHC. Actually, at the pH that is prevalent in these endosomes class I molecules dissociate with detachment of $\beta 2$ microglobulin (27), a condition that may facilitate peptide exchange, afterwards stabilized as the endosomes (or the mini-vesicles derived from them) are neutralized as they return to the plasma membrane. In short, we speculate that activated T cells possess an additional mechanism that is specially adapted to endow their membrane with a display of complexes of TCR peptides and class I MHC (non-classical MHC molecules might be included in this set) that are suitable for recognition by the α/β TCR of CD8+ suppressor T cells.

It is difficult to visualize which TCR peptides might be included in this display; this would depend on the proteolytic processes in the T cells and also might be influenced by the class I alleles available. However, in view of the special dynamics of class I MHC that occurs in activated T cells, we may consider likely that complexes of TCR peptides formed with at least one kind of the different class I molecules available would include peptides of the hypervariable sequence of either the α or β chains of the TCR. These peptides, following the principles of the general network theory, would not be covered by self tolerance and therefore should find complementary CD8$^+$ T cells capable of performing a regulatory function on the basis of a T-T idiotypic interaction.

Procedures are now available (28) for the identification of even minute amounts of peptides bound *in vivo* to class I molecules; therefore, the speculations discussed above for the molecular basis of idiotypic interactions between T cells should be easily testable.

ACKNOWLEDGMENT

Work in the author's laboratory reported here has been supported by LIFEGROUP INC., Padova, Italy.

REFERENCES

1. K. Eichman, I. Falk, and K. Rajewsky, *Eur. J. Immunol.* 8:853 (1978).
2. R. Woodland and H. Cantol, *Eur. J. Immunol.* 8:600 (1978).
3. L. A. Herzenberg, S. J. Black, and L. A. Herzenberg, *Eur. J. Immunol.* 10:1 (1980).
4. I. R. Cohen and H. Atlan, *J. Autoimmunity* 2:613 (1989).
5. A. Bandeira, A. Coutinho, A. Marcos, M. Toribio, and A. C. Martinez, *Ann. Rev. Immunol.* 7:200 (1989).
6. D. Tse, M. Al-Haideri, B. Pernis, C. R. Cantor, and C. Y. Wang, *Science* 234:748 (1986).
7. P. Kourilsky, G. Chaouat, C. Rabourdin-Combe, and J. M. Claverie, *Proc. Natl. Acad. Sci. USA* 84:3400 (1987).
8. J. R. Batchelor, G. Lombardi, and R. I. Leehler, *Immunol. Today* 10:37 (1989).
9. M. P. Happ and E. Heber-Katz, *J. Exp. Med.* 167:502 (1988).
10. H. Acha-Orbea, D. J. Mitchell, L. Timmermann, D. C. Wraith, G. S. Tausch, M. K. Waldor, S. S. Zamvil, H. O. McDevitt, and L. Steinman, *Cell* 54:263 (1988).
11. A. Ben-Nun, H. Wekerle, and I. Cohen, *Nature* 292:60 (1981).
12. O. Liden, T. Reshef, E. Berand, A. Ben-Nun, and I. Cohen, *Science* 239:181 (1988).
13. D. Sun, Y. Qim, J. Chlube, J. T. Epplen, and H. Wekerle, *Nature* 332:843 (1988).
14. A. A. Vanderbark, G. Hashim, and H. Offner, *Nature* 341:541 (1989).
15. M. D. Howell, S. T. Winters, T. Olea, H. C. Powell, D. J. Carlo, and S. W. Brostoff, *Science* 245:688 (1989).
16. H. Offner, G. A. Hashim, and A. A. Vanderbark, *Science* 251:430 (1991).
17. H. Jiang, S.-L. Zhang, and B. Pernis, *Science*, in press (1992).
18. D. R. Koh, F. L. Wai-Ping, A. Ho, G. Dawn, H. Orbea, and T. W. Mak, *Science*, in press (1992).
19. D. Faustman, X. Li, H. Y. Lin, Y. Fu, G. Eisenbarth, J. Avruch, and J. Guo, *Science* 254:1756 (1991).
20. H. Jiang, E. Sercarz, D. Nitecki, and B. Pernis, *New York Acad. Sci.* 636:28 (1991).
21. Y. Minami, A. M. Weissmna, L. E. Samelson, and R. Klausner, *Proc. Natl. Acad. Sci. USA* 84:2688 (1987).
22. C. Chen, J. S. Bonifacino, L. C. Yuan, and R. Klausner, *J. Cell Biol.* 107:2149 (1988).

23. J. S. Bonifacino, J. S. Suzuki, J. Lippincot-Schwartz, A. M. Weissman, and R. Klausner, *J. Cell Biol.* 109:73 (1989).
24. D. Tse and B. Pernis, *J. Exp. Med.* 159:193 (1984).
25. D. Tse, C. R. Cantor, J. McDowell, and B. Pernis, *J. Mol. Cell. Immunol.* 2:315 (1986).
26. J. W. Yewdell and J. R. Benninck, *Adv. Immunol.* in press (1992).
27. J. H. Hochman, H. Jiang, L. Matyus, M. Edidin, and B. Pernis, *J. Immunol.* 146:1862 (1991).
28. D. S. Hunt *et al., Science* 255:1261 (1992).

12

A NETWORK OF SELF INTERACTIONS

Maurizio Zanetti

Department of Medicine and Cancer Center
University of California, San Diego
San Diego, CA 92103

INTRODUCTION

Contemporary Immunology has evolved from two basic paradigms, the *Clonal Selection Theory* and the *Network Theory* which identified clonal expansion by antigen of cells with recognition structures and the regulatory property Ig V region (and possibly TcR V region) antigenic determinants as the main aspects of a functioning, mature immune system. How B and T cells communicate and influence each other is still partially understood.

I recapitulate herein the main points discussed in the two lectures presented at this Course. One is the idiotypic network in its conventional interpretation, and the other new findings related to CD4 and CD45 that makes them part of a network of self interactions. A modern view of the immune system and its physiological functioning may require an integrated approach.

IDIOTYPES OF AUTOANTIBODIES AND SELF-REACTIVE T CELLS

I shall begin by describing classical idiotypy and briefly review our current understanding of it in relation to self reactivity. Idiotypy was discovered in 1963 by Oudin (1) and Kunkel (2) but anti-antibodies that very much resemble anti-idiotypes were observed as early as 1898 by Kossel and Conner and subsequently by Ehrlich and Morgenroth (3). Thus, the idea that the body could normally make auto-antibodies against its own immunoglobulin is as old as modern Immunology. The seventies and the eighties were an exciting period for idiotype research to the point that from the original theoretical view by Jerne (4) many experimental demonstrations came to show the involvement of idiotypy in immune regulation. The collective evidence can be summarized as follows: a) Antibodies (Ab1) produced in response to a given antigen elicits anti-idiotype antibodies (Ab2) in heterologous, homologous, syngeneic and autologous system (5-9); b) Antibodies of the various classes of immunoglobulins in response to the same antigen share idiotypy; c) Anti-idiotypic antibodies arise spontaneously; d) The mutual recognition of idiotype and complementary anti-idiotype occurs both as free molecules and/or cellular receptors on B and T lymphocytes (10).

Idiotypy is defined serologically, *i.e.*, by the reactivity of a polyclonal or monoclonal Ab2 with the variable (V) region of Ab1. Idiotypic determinants can be located in the heavy (H) chain, light (L) chain, or both chains, with most idiotypes mapped to date being localized in the complementarity determining regions (CDRs) of VH and/or VL. Idiotypic determinants can lie outside, close or within the antigen-combining site of the antibody molecule. Therefore, while the interaction between Ab1 and Ab2 may occur at different sites, idiotype detection systems strictly depends upon the type of steric interaction between the variable regions of the two molecules.

Because idiotypes are antigens of the inside the best verification for the idiotype network hypothesis needed to come from studying idiotypy of self reactive antibodies and TcR of autoreactive T cells. I will review idiotypy of self-reactive antibody first.

Over the years many idiotypes have been defined on autoantibodies, both in humans and in rodents. The big aim of these studies was to uncover the existence of molecular and, possibly, genetic relatedness among antibodies with specificity for the same autoantigen. Moreover, the possibility of identifying molecular structures associated with a particular self reactivity or autoimmune disease was regarded as a new way to understand autologous regulation and develop rationale therapies.

In a few cases it has been possible to establish a correlation between a given CRI and the activity of the disease. For instance, in systemic lupus erythematosus (SLE) one idiotype, 16/6, is found in the serum of about 50% of patients with clinically active disease and the level of its expression seems to parallel disease activity (11). A lupus CRI studied by Hanh *et al.* (12), IdGN2, was found to account for the majority of anti-DNA antibodies deposited in the glomeruli of SLE nephritis. In scleroderma individual idiotypes of anti-centromere and anti-topoisomerase autoantibodies in scleroderma appears to correlate with the levels of autoantibody activity (13, 14).

Table I lists the major cross-reacting idiotype (CRI) systems described so far in relation to both systemic and organ-specific autoimmune diseases.

It is important to realize that in the dynamics of the idiotype network auto-Ab2 are also produced. Indeed, anti-idiotypic antibodies specific for autoantibodies have been documented in various autoimmune systems (15-23). Ab2s that functionally mimic the autoantigen at the level of the appropriate somatic or immune cell receptor have also been identified (22-27).

Somewhat similar considerations can be made for the antigen receptor on T cells. Although the available information is still limited it appears that T cells involved in the pathogenesis of autoimmunity display restricted V region gene usage and TcR idiotypes are cross-reacting. For instance, in experimental allergic encephalomyelitis (EAE) T cells that transfer diseases bear receptors whose V regions are highly homologous both in mice and rats (28-30). In humans organ-infiltrating T lymphocytes recognizing

Table I.   CROSS-REACTING IDIOTYPES ON AUTOANTIBODIES

	Autoantibody Specificity	Clinical Association	Rodents	Human
Systemic Diseases	IgG Fc	Rheumatoid Arthritis	(+)	(+)
	Red blood cells	Cold Agglutinin Anemia	(+)	(+)
	DNA	SLE	(+)	(+)
	Sm/RNP	SLE		(+)
	Ro/SSA	SLE/Sjogren		(+)
Organ Specific Diseases	ACh Receptor	*Myasthenia Gravis*	(+)	(+)
	Thyroglobulin	Chronic Thyroiditis	(+)	(+)
	Insulin	Diabetes		(+)
	Myelin Basic Protein	Multiple Sclerosis	(+)	(+)
	Peripheral nerve glycolipid	Neuropathy		(+)
	Kidney Antigens	Interstitial Nephritis	(+)	
	Retinal S antigen	Uveoretinitis	(+)	

immunodominant regions of myelin basic protein (31-33) or lymphocytes from the synovial fluid of patients with rheumatoid arthritis (34, 35) display limited heterogeneity and share TcR Vβ usage.

Regulatory T cells of the Ab2 set with specificity for the TcR of autoimmune effector T lymphocytes have also been documented in the normal as well as in the diseased animal (36-38). It appears therefore, that there exists a parallelism between the B- and T cell repertoire for self antigens with respect to restriction in idiotype expression and V region genes usage. As illustrated in the section below this could constitute the basis for a regulatory network for self-reactive molecules of the immune system.

THE CONCEPT OF AUTOIMMUNE NETWORK

The process underlying the establishment of immunological tolerance is currently viewed as the product of negative selection in the thymus (this will be reviewed elsewhere in this volume). Its maintenance, however, particularly in peripheral lymphoid organs, is still in search for a plausible explanation. While anergy at the T cell level can only be accounted for cells reacting with autoantigens released in the circulation, active idiotype regulation may play an important role as it is independent of antigen. A few years ago I defined the *autoimmune network* as the fundamental, comprehensive framework of self-recognition/reactivity events in which responsiveness to self antigens is modulated through the positive and negative influence of antibody:antibody (humoral) or receptor:anti-receptor (cellular) interactions based on domain/domain interactions between antigenic sites of antibody and TcR V regions (39).

There exists now a large body of experimental evidence that the autoimmune network is a functional reality and possibly one of the important regulatory mechanisms available to the immune system for the maintenance of tolerance vis-a-vis clones that escaped deletion in the thymus or peripheral anergy. Table II recapitulates the existing evidence for a functional autoimmune network.

Table II. EVIDENCE FOR THE EXISTENCE OF AN AUTOIMMUNE NETWORK

1. The humoral response to self antigens can be suppressed by passive administration of antiidiotypic antibodies (40-43).
2. The humoral response to self antigens can be suppressed by active immunization with idiotype (44, 45).
3. The humoral response to self can be suppressed *in vivo* by passive transfer of idiotype-primed T cells (44, 45).
4. Humoral and cellular responses to self antigens can be suppressed *in vivo* by active immunization with antigen- or idiotype-reactive T cells (46 , 47)
5. Active immunity using synthetic peptides of disease-inducing T cell receptors prevent the induction of experimental disease [48,49]

6. Self-reactive antibodies can be induced in naive animals by active immunization with idiotypic (50) and anti idiotypic (51) antibodies.
7. Immunization with a self antigen may induce anti-receptor autoantibodies via an idiotypic mechanism (52-58).

While little doubt exists that there interactions occur in the body it is not yet understood how the various functions integrate themselves at any given time, how suppression is realized (antibody-mediated cytotoxicity, complement -mediated inactivation of self-reactive clones, cytotoxic T cells etc.) and why, in converse, activation is the prevalent force in some instances.

In the course of our studies we found to additional mechanisms whereby the V regions of autoantibodies can directly influence the autoimmune network and cooperate in the maintenance of peripheral tolerance. The first is the spontaneous occurrence of antibodies to T cell membrane structures capable of silencing T cell activation. The other is a new mechanism available to the immune system to increase the immunogenicity of self idiotopes, an interaction between CD4 and Ig V regions.

NATURAL AUTOANTIBODIES TO CD45

A few years ago while studying the B cell repertoire of neonatal Balb/c mice we reported the biological activity of an autoantibody to the common leucocyte antigen CD45 (59). In humans and in rodents CD45 consists of a family of glycoproteins of high molecular weight expressed on most cells of the hematopoietic lineage including B and T cells, monocytes and mastocytes (60). Its expression varies depending on the type and the degree of cell differentiation and activation (61) and it appears to be highly restricted to cellular components of the immune system. It has been suggested that CD45 could play a role in the process of maturation and/or differentiation of immune cells (62), mitogen-induced proliferation of T cells (63), mixed lymphocyte reaction (64), T lymphocyte cytotoxicity (65) and B cell activation (66). While little is known about its natural ligands, recent evidence indicates that its intracellular domains function as a protein-tyrosine phosphatase (67), and is involved in $p56^{lck}$-linked phosphorylation during TcR/CD3 complex-dependent activation. Cross-linking of the intracytoplasmic domain of CD45 with CD4 and/or the CD3 $p56^{lck}$ complex (68) potentiates phosphorylation of the ζ chain and regulates T cell activation. CD45 is also linked with CD2 (69) on alternative receptor that can induce proliferation and differentiation of T lymphocytes.

An autoantibody isolated from the spleen of neonatal mice was able to inhibit (>90% inhibition) the incorporation of ^{3}H-thymidine by T lymphocytes stimulated *in vitro* with concanavalin A (Con A), and blocked the production of interleukin 2 (IL-2) by activated T cells. Although we do not know whether naturally-occurring autoantibodies to CD45 could prevent the activation of T cells by endogenous lectin-like ligands or the MHC/peptide antigen complex, it is tempting to speculate that such autoantibodies could play a role in buffering the activation of T cells early in ontogeny and prevent the expansion of potentially deleterious B-cell clones.

CD4/Ig INTERACTION

Recently, we and others found that CD4 interacts with human and murine immunoglobulins (Ig) (70, 71). We characterized the CD4/Ig binding as a low affinity interaction apparently mediated by residues in the Fd region of the H chain on the Ig side and by residues 21-49 of the first extracellular domain of CD4 on the receptor side. Both CD4 and CD4-derived synthetic peptides corresponding to this region blocked the interaction of Ig with CD4 (70). CD4 also binds Ab complexed with Ag (Ab:Ag complex). The avidity of interaction between CD4 and Ig was estimated approximately 100 fold less than that of CD4 for Ab:Ag complex. Binding of Ab:Ag complex increased progressively as a function of the amount of Ag added. Maximum binding was reached at an Ab:Ag molar ratio characteristic of soluble complex. Table III shows a quantitative analysis of the avidity of interaction between CD4 or synthetic peptide 21-49 and Ab:Ag (anti-DNP:DNP$_{37}$BSA) complexes vs. monomeric or heat-aggregated anti-DNP Ab, indicating that CD4 and CD4 peptide 21-49 bound Ab:Ag complex formed at a molar ratio of 6:1 many folds more effectively than Ab alone.

Table III.   Enhanced binding of Ab:Ag complex to CD4

Wells Coating	Equivalent nM of an anti-DNP antibody			
	Monomeric Ab	Aggregated Ab	Ab:Ag complex	Ab:Ag / Ab Ratio
CD4	156[*]	312	15	~10:1
p16-49	156	156	0.8	~150:1
p39-60 BSA	>10^4	>10^4	>10^4	>10^4

[*] Values refer to equivalent amounts (nM) of Ab required to give an O.D. A$_{492}$ of 0.5.

Subsequently, two additional relevant observations were made. One is that CD4 binds Ab1:Ab2 complex far more efficiently than Ab1 alone (72). The second is that synthetic peptide 21-49 of CD4 increases binding of human (H) Ig to U937 monocytic cells (73) (Table IV).

Table IV.   CD4 peptide 21-49 enhances the binding of A-Ig to U937 cells

	Ligand	Cell-bound		Catabolized		Total uptake	
		cpm	(%)	cpm	(%)	cpm	(%)
H Ig	Medium	772[*]	(3.5)	144	(0.7)	916	(4.2)
"	CD4 p21-49	14,167	(50.9)	95	(0.4)	14,262	(51.3)
"	Control peptide	800	(3.5)	150	(0.8)	950	(4.3)

The site of CD4 involved in enhanced binding was mapped using CD4 peptides truncated either at the amino- or the carboxy-terminal of the 21-49 region, the same CD4 residues involved in the Ig-binding activity were identified.

Ab:Ag complex regulate both cellular and humoral immune responses (74). They interact with antigen receptor (mIg)-bearing B lymphocytes and FcR expressed on B and T cells, monocytes, macrophages and other nonlymphoid cells. Ab:Ag complex may augment or suppress humoral responses mainly by crosslinking FcR on the cell surface. They also exert positive or negative regulation on antibody-dependent cell-mediated toxicity (ADCC), cell-mediated lymphocytic reactions and DTH. The differential regulatory activity, positive or negative, exerted by Ab:Ag complex on the immune response depends, among other factors, on their molar ratio. For example, in Ag excess ADCC is inhibited, whereas in Ab excess lymphocytes are armed and ADCC is enhanced (75, 76). Therefore, a participation of CD4 in Ab:Ag complex uptake may have great implications for cell-to-cell interaction during innate and induced immune responses and the response to idiotype-anti-idiotype complex (77, 78)

It is well known that, at the functional level, Ab:Ag complex elicit anti-idiotypic antibodies to self-idiotopes and antigen-reactive T cells and anti-idiotypic B cells cooperate in the anti-idiotype response to Ab:Ag complex (79, 80). Thus, a potential new role of CD4 in increasing uptake Ab:Ag complex by FcR may lead to greater internalization, transport to the lysosomal compartment, chemical degradation and ultimately presentation to TcR. This could , therefore, be a new way to enhance the immunogenicity of self idiotype peptides and increase regulation during the early phase of Ab:Ag complex formation.

CONCLUSIONS

I sketched above three possible scenarios for the interaction between self structures. These included members of the Ig gene superfamily (Ig, TcR and CD4) and CD45, a molecule highly represented at the surface of lymphocytes and functionally implicated in their activation. While the idiotype network remains the main paradigm for immune regulation the experiments and the arguments presented suggest that the original Jernian concept may need to be revised to encompass other molecular interactions between B-T cells. It is certainly easy to envision how the immune system, particularly during early ontogeny, could take advantage from the anti-T cell activity of natural anti-CD45 autoantibodies. It is also becoming clearer what the role might be for a CD4/Ig interaction in the overall economy of immune regulation. The mere fact that Ab alone bind with higher avidity than Ab alone point out a regulatory role by immune complex, be they of the Ab:Ag or Ab1:Ab2 type. In both cases increased binding/internalization that may follow capture of immune complex by CD4, could help to further concentrate the regulatory structures of the idiotype network (Ab1 or Ab2) on immunocompetent cells.

This event would be most relevant when Ab1 and/or Ab2 are present in the serum at low concentrations and in soluble form. This focussing device could bring a new dimension to the way we look at immune regulation. In conclusion, self structures are embedded in a highly articulated network of interactions whose physical nature needs further study and understanding.

ACKNOWLEDGEMENTS

This work was supported in part by a grant from National Institute of Health (HD 25787)

LITERATURE CITED

1. J. Oudin and M. Michel., *C.R. Acad. Sci. Ser. D* 257, 805-808 (1963).
2. H. G. Kunkel, M. Mannik, R. C. William, 140, 1218-1219 (1963).
3. A. M. Silverstein, *A History of Immunology* (Academic Press Inc., San Diego, 1989).
4. N. K. Jerne, *Ann. Immunol. (Paris)* 125, 373-389 (1974).
5. M. G. Kuettner, A. L. Wang, A. Nisonoff, *J Exp Med* 135, 579-595 (1972).
6. J. Oudin, *Ann. Immunol. (Paris)* 126C, 309-337 (1974).
7. L. S. Rodkey, *J Exp Med* 139, 712-720 (1974).
8. J. Roland, P.-A. Cazenave, *Eur J Immunol* 11, 469-474 (1981).
9. N. Sakato, H. N. Eisen, *J Exp Med* 141, 1411-1426 (1975).
10. K. Eichmann, *Adv Immunol* 26, 195-254 (1978).
11. D. Isenberg, *et al.*, *J. immunol.* 135, 261-264 (1985).
12. B. H. Hahn, *et al.*, *Arthritis Rheum* 33, 978-984 (1990).
13. S. Hildebrandt, E. Weiner, W. Earnshaw, M. Zanetti, N. Rothfield, (1991).
14. S. Hildebrandt, *et al.*, 10, 41-48 (1991).
15. N. I. Abdou, H. Wall, H. B. Lindsley, J. F. Halsey, T. Suzuki, *J Clin Invest* 67, 1297-1304 (1981).
16. P. L. Cohen, R. A. Eisenberg, *J Exp Med* 156, 173-180 (1982).
17. D. S. Dwyer, R. J. Bradley, C. K. Urquhart, J. F. Kearney, *Nature.* 301, 611-614 (1983).
18. A. K. Lefvert, R. W. James, C. Alloid, B. W. Fulpius, *Eur. J. Immunol.* 12, 790-792 (1982).
19. T. Abe, *et al.*, *J Immunol* 132, 2381-2385 (1984).
20. M. Zouali, J. M. Fine, A. Eyquem, *J Immunol* 133, 190-194 (1984).
21. D. Elias, I. R. Cohen, Y. Schechter, Z. Spirer, A. Golander, *Diabetes* 36, 348-354 (1987).
22. W. K. Hancock, E. V. Barnett, *Clin Exp Immunol* 75, 25-29 (1989).
23. R. S. Geha, *et al.*, *N Engl J Med* 312, 534-540 (1985).
24. S. Fong, T. A. Gilbertson, D. A. Carson, *J Immunol* 131, 719-723 (1983).
25. T. L. Moore, T. G. Osborn, R. W. Dorner, *Arthritis Rheum* 32, 699-705 (1989).
26. H. Eng, A. K. Lefvert, *Ann Inst Pasteur Immunol* 139, 569-580 (1988).
27. H. M. Sikorska, *J Immunol* 137, 3786-3795 (1986).
28. H. Acha-Orbea, *et al.*, 54, 263 (1988).
29. J. Urban, *et al.*, 54, 577-592 (1988).
30. E. Heber-Katz, H. Acha-Orbea, *Immunol. Today* 10, 164-169 (1989).
31. B. L. Kotzin, *et al.*, *Proc Natl Acad Sci* 88, 9161-9165 (1991).
32. J. R. Oksenberg, *et al.*, *Nature* 345, 344-346 (1990).
33. K. W. Wucherpfennig, *et al.*, *Science* 248, 1016-1019 (1990).
34. X. Paliard, *et al.*, *Science* 253, 325-329 (1991).
35. M. D. Howell, *et al.*, *Proc Natl Acad Sci* (1991).
36. D. H. Adda, E. Beraud, R. Depieds, *Eur. J. Immunol.* 7, 620-625 (1977).
37. E. Beraud, S. Varriale, C. Farnarier, D. Bernard, *Eur. J. Immunol.* 12, 926-930 (1982).

38. A. Ben-Nun, Y. Ron, I. R. Cohen, *Nature* 288, 389-390 (1980).
39. M. Zanetti, D. Glotz, J. Rogers, K. Meek, C. Hasemann, *Monogr Allergy* 22, 46-56 (1987).
40. M. Zanetti, P. E. Bigazzi, *Eur. J. Immunol.* 11, 187-195 (1981).
41. B. H. Hahn, F. M. Ebling, *J Immunol* 132, 187-190 (1984).
42. M. Zanetti, C. B. Wilson, *J Immunol* 130, 2173-2179 (1983).
43. M. Zanetti, F. Mampaso, C. B. Wilson, *J. Immunol.* 131, 1268-1273 (1983).
44. E. G. Neilson and M. S. Phillips., *J. Exp. Med.* 155, 179-189 (1982).
45. M. Zanetti, D. Glotz, J. Rogers, *J Immunol* 137, 3140-3146 (1986).
46. E. G. Neilson, E. McCafferty, S. M. Phillips, M. D. Clayman, C. J. Kelly, *J Exp Med* 159, 1009-1026 (1984).
47. D. B. Wilson, *Immunol Today* 5, 228-230 (1984).
48. M. D. Howell, *et al.*, *Proc Natl Acad Sci* 88, 10921-10925 (1991).
49. A. A. Vandenbark, G. Hashim, H. Offner, *Nature* 341, 541-544 (1989).
50. T. L. Hall, R. B. Colvin, K. Carey, R. T. McClusky, *J Exp Med* 146, 1246-1260 (1977).
51. M. Zanetti, J. Rogers, D. H. Katz, in *Regulation of the immune system* C. Secartz Chess, Eds. (Liss, New York, 1984), pp. 893-907.
52. K. Sege and P. A. Peterson., *Proc. Natl. Acad. Sci. USA* 75, 2443-2447 (1978).
53. Y. Shechter, R. Maron, D. Elias, I. R. Cohen, *Science* 216, 542-545 (1982).
54. N. H. Wasserman, *et al.*, *Proc Natl Acad Sci USA* 79, 4810-4814 (1983).
55. W. L. Cleveland, N. H. Wasserman, R. Sarangarajan, A. S. Penn, B. F. Erlanger, 305, 56-57 (1983).
56. M. N. Islam B. M. Pepper, R. Briones-Urbina, and N. R. Farid., *Eur. J. Immunol.* 13, 57-63 (1983).
57. J. R. Baker, Y. G. Lukes, K. D. Burman, *J. Clin. Invest.* 74, 488-495 (1984).
58. A. B. Schreiber, P. O. Couraud, C. Andre, B. Vray, A. D. Strosberg, *Proc. Natl. Acad. Sci. USA* 77, 7385-7389 (1980).
59. D. Glotz, M. Zanetti, *J. Immunol.* 142, 439-443 (1989).
60. S. Shaw, *Immunol. Today* 8, 1-3 (1987).
61. L. Lefrancois, M. J. Bevan, *J. Immunol.* 135, 374-383 (1984).
62. L. Lefrancois, L. Puddington, C. E. Machamer, M. J. Bevan, *J. Exp. Med.* 162, 1275-1293 (1985).
63. J. A. Ledbetter, *et al.*, *J. Immunol.* 135, 1819-1825 (1985).
64. J. A. Harp, S. J. Ewald, *Cell Immunol.* 81, 71-80 (1983).
65. J. C. Weill, C. A. Reynaud, *Science* 238, 10940-1098 (1987).
66. R. S. Mittler, R. S. Greenfield, B. Z. Schacter, N. F. Richard, M. K. Hoffmann, *J. Immunol.* 138, 3159-3166 (1987).
67. N. K. Tonks, C. C. Diltz, E. H. Fischer, *J. Biolog. Chem.* 265, 106-174 (1990).
68. C. E. Rudd, P. Anderson, C. Morimoto et al., *Immunological Rev.* 111, 225 (1989).
69. B. Schraven, Y. Samstag, P. Altevogt et al., *Nature* 345, 71-74 (1990).
70. P. Lenert, D. Kroon, H. Spiegelberg, E. S. Golub, M. Zanetti, *Science* 248, 1639-1643 (1990).
71. S. Lederman, M. J. Yellin, A. M. Cleary, R. Gulick, L. Chess, *J. Immunol.* 144, 214-220 (1990).
72. P. Lenert, M. Zanetti, *Inter. Rev. Immunol.* 7, 237-244 (1991).
73. P. Lenert, R. L. Mehta, M. Zanetti, *J Immunol* in press (1992).

74. A. Theophilopoulos, F. J. Dixon, *Adv. Immunol.* 28, 89-220 (1979).
75. H. J. Lustig, C. Bianco, *J. Immunol.* 253-260 (1976).
76. E. Saksela, T. Imir O., *J. Immunol.* 115, 1488-1492 (1975).
77. A. J. Morgan, R. D. Rossen, J. J. Twomey, *J Immunol* 122, 1672-1680 (1979).
78. D. Geltner, E. C. Franklin, B. Frangione, *J Immunol* 125, 1530-1535 (1980).
79. G. G. Klaus, *Adv Exp Med Biol* 114, 289-294 (1979).
80. G. G. Klaus, *Nature* 278, 354-355 (1979).

13

STIMULATION OF LYMPHOCYTES BY ANTI-IDIOTYPES BEARING THE

INTERNAL IMAGE OF VIRAL ANTIGENS

Habib Zaghouani and Constantin Bona

Department of Microbiology,
Mount Sinai School of Medicine
One Gustave Levy Place
New York   NY 10029

INTRODUCTION

Immunoglobulin (Ig) and T cell receptor (TCR) variable regions exhibit dual properties of recognizing antigens and being antigenic by virtue of expressing idiotypes. As for antigens which can be made of a single antigenic determinant (i.e epitope) or a set of epitopes, an idiotype can be made of a single or a set of idiotopes. According to the idiotype network theory[1] each epitope in the universe can be mimicked by an idiotope. In other words idiotopes are " internal images " of epitopes. Based on this postulate, the duality of Ig and the large repertoire of variable regions, it has been suggested that the immune system is in a steady state of interactions involving the combining site or paratopes and the idiotopes of variable regions. Introduction of an antigen into this system perturbs such a steady state leading to a readjusted equilibrium.

As stated above, variable regions of Ig and TCR specific for a given epitope express an idiotope. Introduction of such a variable region into the immune system would then activate clones through both its combining site and its idiotope[2]. The clones that are specific for the idiotope can be of Ab2α or Ab2γ type[3] and those induced through the combining site are of Ab2β type[4]. The latter are also called internal images because they bear an idiotope that is able to mimic such an epitope[4].

Ab2β antibodies are able to mimic antigens and stimulate the specific clones in-vitro as well as in-vivo[5-8]. This is of particular interest in the development of vaccines because Ig

T Lymphocytes: Structure, Function, Choices, Edited by F. Celada
and B. Pernis, Plenum Press, New York, 1992

are self molecules devoid of side effects compared to microbial vaccines and have a higher half life compared to synthetic peptides. Herein, we discuss further the heterogeneity of anti-idiotypic (anti-id) antibodies, the criteria to define Ab2ß and suggest a new strategy to generate antibody of Ab2ß type and provide evidence that these antibodies are able to be recognized by either B or T cells.

Heterogeneity of Anti-Idiotypic Lymphocytes

Idiotypes borne by the receptors of either B or T cells were initially defined by serological reagents called anti-idiotypic antibodies[9]. Later, it was shown that there is a perfect symmetry in the immune system because Ab2 clones recognizing idiotypes were found at the level of both B and T cells[10]. B lymphocyte clones produce Ab2 antibodies that recognize idiotypic determinants expressed on receptors of either B or T cells[11,12]. These Ab2 antibodies can be specific for either an idiotope shared by various clones (cross reactive idiotope) or an individual idiotope that can be found on the receptor of a single clone[13]. Similarly T cell clones specific for idiotopes borne by Ig or TCR from the CD8 and CD4 populations have been defined[14,15]. Because T cells recognize peptides in association with MHC antigens, one would expect that Ig carrying T cell idiotopes would have to be processed in order to be able to mediate specific T cell recognition[16]. By contrast it is not known whether T cell clones that are specific for an idiotope borne by TCR bind directly to the cell surface receptor or also requires the processing of TCR and generation of the idiopeptide.
So far, based on functional and structural properties, four major categories of Ab2 have been defined:
Ab2α consists of a subset of clones expressing receptors able to recognize framework-associated idiotopes borne by Ig or the TCR of other clones. The product of such clones can also interact with regulatory idiotopes[17], which are expressed on the receptors of clones specific for different epitopes[18].
Ab2γ consists of a subset of clones producing antibodies against idiotopes associated with the combining sites which are inhibitable by the antigen[3].
Ab2ß consists of a subset of clones expressing receptors that bear idiotopes that resemble the antigen. Through molecular mimicry, the products of such clones can stimulate antigen specific clones[4].
Ab2ε consists of a subset of clones expressing receptors that can interact with epitopes as well as with idiotopes. The products of such clones are designated epibodies[19].
Numerous data demonstrated that Ab2 clones have regulatory effects on immune responses and that they are able either to enhance or suppress specific responses depending on the amount used, their affinity and origin[20].

Since Ab2ß antibodies are able to elicit specific humoral response in lieu of antigen, they represent good candidates for vaccines. However, the high heterogeneity of Ab2 antibodies the difficulties encountered to distinguish Ab2ß among other Ab2s and the low frequency at which they occur have limited their usage as vaccines. To circumvent these practical difficulties several criteria were proposed to define an antibody as Ab2ß.

Criteria to Define an Ab2 as Internal Image

It appears that the differences between Ab2ε and other Ab2s are straight forward. However, it is much more difficult to distinguish Ab2ß from Ab2α and Ab2γ. Recently, the following four criteria have been proposed to define an antibody as Ab2ß: 1) the interaction between Ab2ß and the antigen specific receptor should be inhibitable by the antigen 2) Ab2ß should bind to receptors specific for the same antigen but raised in different animal species. 3) the majority of the clones that would be stimulated in-vivo in a syngeneic system would be specific for the original epitope independent of their idiotopes. 4) Finally, a sharing of similar or identical sequence structure by the variable regions of Ab2ß and the original antigen is the best criterion that define an Ab2ß[16]. The first Ab2ß fulfilling these was described by Sege and Peterson[21]. In these experiments an anti-idiotope against rat antibodies specific for bovine insulin bound to insulin receptor of fat cells and mimicked the effect of insulin. Following this observation a legion of experimental data demonstrated that Ab2ß can mimic microbial proteins, polysaccharides, steroid hormones, drugs and other biologically active ligands[22].

In the case of Ab2ß carrying the internal image of T cell epitopes, one would predict that they must be processed to generate an "idiopeptide" that would associate with MHC antigens in order to stimulate the T cells. As stated above the most important practical issue for investigators involved in the utilization of Ab2ß as antigen surrogates is to distinguish them form other Ab2 reagents. We have recently shown that an Ab2 antibody specific for the envelope protein of the human immunodeficiency virus can indeed bind envelope specific antibodies raised in various animal species or obtained from an individual infected with HIV (see below). Shared amino acids between Ab2ß and the original antigen has been described for the first time by Fougereau and colleagues[23]. In this case it was shown that an Ab2 antibody that was able to stimulate clones specific for Glu-Tyr polymer bears Glu-Tyr-Tyr tripeptide within the D segment of the heavy chain variable region. Similarly, it was shown that an Ab2 antibody raised against antibodies specific for the hemagglutinin of reovirus shared within the CDR2 of its light chain variable region a tetrapeptide that is identical to a stretch of amino acids corresponding to a neutralizing epitope of the hemagglutinin of reovirus[24]. Obviously this criterion can not be used for Ab2ß which mimic non-protein epitopes. In this case crystallography analysis might shed light on whether these antibodies have contacting residues that display similar charges as the antigen. It would be then of absolute necessity to assess the response that can be induced in-vivo. If the Ab2 is of the ß type the immune response that would be induced in a syngeneic system would be dominated by clones that are reactive to the nominal antigen[25].

RESULTS

Characterization of Anti-Idiotypes Raised Against Antibodies Specific for a B Cell Epitope of The Envelope Protein of The Human Immunodeficiency Virus

The amino acid residues 503-535 of the envelope protein of HIV represent an immunodominant B cell epitope able to induce virus specific antibodies in HIV infected individuals[26]. A peptide encompassing these amino acid residues was synthesized, coupled to keyhole limpet hemocyanin (KLH) and used to immunize rabbits, baboons and mice. Anti-503-535 antibodies were purified from the serum of these animals and tested for binding to the synthetic peptide and to the envelope protein by radioimmunoassay. As can be seen in Table 1 while the rabbit and mice antibodies bind to both 503-535 and to gp160 envelope protein, the baboon antibodies bind weakly to the peptide and poorly to the envelope protein. However, by radioimmunoprecipitation using ten ug of affinity purified antibodies we obtained precipitation of the envelope protein with rabbit, baboon and mouse antibodies as well as with antibodies purified from an individual infected with HIV[27]. The immune serum of rabbits and mice were able to neutralize both HIV-1 IIIB and MN isolates[28].
In further experiments, mice were immunized with the rabbit anti-peptide antibodies and used to generate monoclonal anti-idiotypes. Three antibodies of IgG1,K isotypes were selected and tested for binding to anti-503-535 antibodies obtained from various species. As can be seen in Table 2. the monoclonal antibodies bind strongly to rabbit anti-peptide antibodies used to immunize the mice but do not bind to normal immunoglobulin of the same rabbit. These antibodies bind also to anti-503-535 antibodies obtained from baboons, mice and human individual infected with HIV. These latter bound weakly compared to rabbit antibodies but significantly higher than normal Ig from the same animal.
As stated above the binding of anti-idiotypes to antibodies specific for the same epitope but obtained from various animal species would indicate that these anti-idiotypes bear an antigenic determinant that is able to mimic the original antigen rather than recognizing a common idiotope borne by the antibodies from various species. Furthermore, we have shown that the binding of anti-idiotypes to rabbit anti-peptide antibodies is inhibited by the synthetic peptide coupled to BSA[27]. Collectively, these data support the idea that the anti-idiotypes are of the ß type.

Furthermore rabbits sensitized with the monoclonal antibodies produced anti-peptide antibodies that were able to recognize the envelope protein of HIV. However, because of possible induction of anti-mouse isotypic and allotypic antibodies we were not able to analyze the heterogeneity of the immune response specific for the variable region of the anti-idiotypes. Based on in-vitro studies of the interaction of anti-idiotypes with the anti-503-535 antibodies we believe that they are of Ab2ß type but a definitive conclusion on this possibility requires more thorough analysis.

Table 1. Binding of purified anti-peptide antibodies to 503-535 peptide and to the envelope protein of HIV

	BSA	735-752	503-535	gp160
rabbit	114 ± 25	81 ± 61	15546 ± 332	2552 ± 61
mice	175 ± 15	237 ± 13	8812 ± 139	1860 ± 110
baboon	107 ± 12	103 ± 14	2832 ± 75	380 ± 51

Microtiter plates were coated with the various antigens, saturated with goat serum, and incubated with 10 ng of each of the purified anti-503-535 antibodies. Bound antibodies were revealed with iodine labeled anti-rabbit Ig, anti-mice Ig and anti-human Ig for rabbit, mice and baboon anti-peptite antibodies respectively.

Table 2. Binding of monoclonal anti-idiotypes to anti-peptide antibodies of various animal species.

Binding to	3B10	4C8	5A12
Rabbit			
anti-505-535	27618 ± 394	17339 ± 349	12864 ± 185
preimmune Ig	354 ± 24	413 ± 26	307 ± 11
Murine			
anti-503-535	1442 ± 57	1963 ± 156	1707 ± 77
preimmune Ig	412 ± 18	452 ± 22	276 ± 12
Baboon			
anti-503-535	726 ± 57	814 ± 26	926 ± 14
preimmune Ig	306 ± 26	391 ± 14	322 ± 27
Human			
anti-503-535	1023 ± 105	1594 ± 213	909 ± 15
normal Ig	324 ± 19	279 ± 123	240 ± 12

The binding of anti-idiotype monoclonal antibodies to anti-peptide antibodies was carried out by radioimmunoassay in which plates were coated with affinity purified anti-503-535 antibodies, saturated with 2% BSA and incubated with ^{125}I-labeled anti-idiotype monoclonal antibodies.

The difficulties encountered in producing Ab2ß antibodies is related to the low frequency of such clones in the repertoire and to the uncertainty of distinguishing Ab2ß from other Ab2s. Thus, the conventional methods used to produce Ab2ß antibodies are not attractive at least in the case of protein antigens. An alternative strategy to prepare Ab2ß anti-idiotypes relies on molecular biological methods which permits the cloning and the expression of Ig genes. In fact, in the case of defined B or T cell epitopes of protein nature one can express these epitopes on an Ig molecule and prepare Ab2ß antibodies on the basis of the most reliable structural criterion used to define an anti-idiotype as internal image of antigen. An important requirement which should be taken into consideration is that the foreign epitope should be surface exposed and accessible for interaction with corresponding clones. In addition, the insertion of the foreign epitope into an Ig molecule should not alter the heavy and light chain pairing. It should be also mentioned that this strategy has the advantage of circumventing the low frequency of Ab2ß precursors, a limiting factor of the conventional strategy for preparing Ab2ß antibodies.

Engineering of Anti-Idiotypic Antibodies Bearing T Cell Epitopes on Their Heavy Chain Variable Regions

In order to test the strategy to prepare anti-idiotypic antibodies carrying T cell epitopes, we used the heavy and light chain genes encoding the anti-arsenate antibody,91A3[29,30]. Because idiotypes are antigenic determinants of variable regions, we choose to express epitopes within the heavy chain variable region. The two epitopes that were selected for testing this strategy correspond to the cytotoxic and helper T cell epitopes of influenza virus nucleoprotein and hemagglutinin, respectively. The cytotoxic T cell (CTL) epitope is 15 amino acid long and correspond to residues 147-161 of influenza virus nucleoprotein. This epitope has been shown to be recognized by CTLs in association with K^d MHC class I antigen[31]. The helper epitope is 11 amino acid long and correspond to residues 110-120 of influenza virus hemagglutinin. This epitope is recognized by T helper cell in association with $I-E^d$ MHC class II antigen[32].

As can be seen in figure 1. a 5.5 kb DNA fragment encompassing the VDJ coding sequence of 91A3 heavy chain and the necessary regulatory elements was used in a polymerase chain reaction to delete the D segment and insert instead an oligonucleotide that encodes either the CTL or helper T cell epitope[33]. We choose to insert the oligonucleotide in place of the D segment because this region is variable in length and represents the third complementary determining region of the heavy chain which is known to be exposed and accessible for interaction with arsenate hapten. These fragments which carry separately the nucleoprotein and hemagglutinin epitopes were subcloned into a pSV2 expression vector[34] upstream of the exons of a Balb/c γ2b constant region gene. In order to produce a complete 91A3 immunoglobulin we transfected the Balb/c non-secreting myeloma B cell line with a vector carrying the entire 91A3 light chain gene[30] and the vector carrying the

heavy chain gene. After selection, we obtained cells that
secrete 91A3 antibody bearing the nucleoprotein epitope in
place of the D region (designated 91A3-NP) as well as cells
which produce antibodies bearing the hemagglutinin T helper
epitope (designated 91A3-HA). We also transfected cells with
the plasmid carrying the light chain together with a plasmid
carrying unmodified heavy chain. The antibody obtained from
these cells (designated 91A3) does not carry any viral epitope
and was used as control in various experiments.
In a preliminary experiment we tested the 91A3 transfectomas
with respect to production of complete immunoglobulin
structure, binding to antibodies specific for a synthetic
peptide (NP-peptide) corresponding to residues 147-161 of
influenza virus nucleoprotein and for binding to arsonate
hapten. As illustrated in table 3, both the Ig bearing the

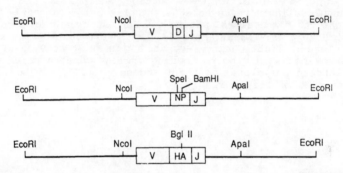

Figure 1. Replacement of the 91A3 heavy chain D segment
by T cell epitopes

CTL epitope and the wild type Ig captured by a rat anti-mouse
K light chain antibody bind to goat anti-mouse Fc IgG2b
antibodies. These results suggest that the insertion of the
CTL epitope in place of the D region did not alter the heavy
and light chain pairing. Furthermore, rabbit antibodies
specific for the peptide of the nucleoprotein bind to 91A3
expressing the CTL epitope but did not bind the 91A3 wild type
antibody.
Subsequently, we tested the antibody for binding to arsonate.
As can be seen in figure 2, both 91A3 antibody produced by
transfectoma cell and 91A3 antibody produced by hybridomas
cell bind to arsonate but the 91A3-NP lost the arsonate
binding activity. These results are in good agreement with the
postulate that the D region is crucial for arsonate binding[35].

In order to test the ability of 91A3 antibody and the
transfectoma cell producing this antibody for mediating T cell
recognition, we generated a CTL line specific for synthetic NP

Table 3. Immunochemical properties of 91A3-NP antibody expressing the nucleoprotein CTL epitope.

Binding to[*]	antibody	
	91A3	91A3-NP
rat anti-mouse K	42724 ± 127[+]	59431 ± 349
rabbit anti-NP	5616 ± 217	1246 ± 76

[*] The immunoglobulin bearing the NP epitope (91A3-NP) was tested along with the wild type 91A3 antibody for binding to rabbit anti-NP peptide by radioimmunoassay in which a 1/3 dilution of transfectoma supernatant was incubated on plates coated with rabbit anti-NP peptide antibodies and bound immunoglobulins were revealed with ^{125}I-labeled rat anti-mouse K light chain monoclonal antibody. For assessment of complete Ig structure of 91A3-NP antibody plates were coated with a rat anti-mouse K light chain, incubated with 10 ng of 91A3-NP and 91A3 antibodies and bound immunoglobulins were revealed with ^{125}I-labeled polyclonal goat anti-mouse Fc IgG2b antibodies.
[+] Mean ± SD of triplicates.

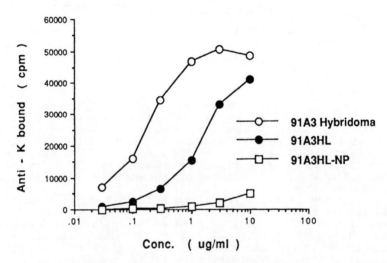

Figure 2. Loss of anti-arsenate binding activity of 91A3HL-NP antibody

peptide by immunizing Balb/c mice with PR8 influenza virus and stimulating splenic cells with synthetic peptide in-vitro. In a preliminary experiment we found that this CTL line is able to lyse SP2/0 cells infected with PR8 virus or coated with NP peptide but does not lyse cells infected with different subtype of influenza virus or coated with irrelevant peptide. We then used this NP specific CTL to assess recognition by T cells of both the transfectomas cells producing the 91A3-NP antibody and SP2/0 cell pulsed with soluble 91A3 antibody. As can be seen in Table 4 transfectoma cells expressing the 91A3 heavy chain carrying the NP epitope and the wild type light chain were lysed by NP specific CTL like SP2/0 cells coated with the NP peptide, while transfectoma cells expressing the wild type heavy and light chains genes as well as SP2/0 cells coated with an irrelevant peptide were not. These data indicate that the heavy chain carrying the NP epitope is processed and an NP peptide is generated which is able to associate with class I K^d antigen and mediate recognition by the NP specific CTL. Furthermore, the cytolysis of transfectoma sxpressing the NP epitope within the heavy chain by NP-specific CTL was inhibitable by anti-K^d but not an anti-D^b monoclonal antibodies. On the other hand SP2/0 cells pulsed with soluble 91A3-NP antibody were not lysed by the NP specific CTL (Table 5) while SP2/0 cells coated with NP peptide were lysed.

The overall conclusion of these experiments is that the 91A3-NP is true Ab2ß since it bears the NP epitope which is recognized by rabbit anti-NP peptide antibodies. The cells expressing 91A3-NP genes generate the peptide and present it in association with class I antigen to a specific CTL. However the Ab2ß 91A3-NP antibody can not mediate recognition by CTL. This is in good agreement with the generally accepted concept that only the endogenous pathway is able to generate peptide that associate with class I MHC antigen and recognized by CD8[+]-T cells. In other words, antigens that are synthesized by antigen presenting cells (APC) themselves are proteolysed in cytosolic compartment to generate peptides which associate with MHC class I antigens and are recognized by CD8[+]-T cells[36,37]. By contrast, the antigens that are exogenously added to APC are taken up by these APC and proteolysed in the endosomal compartment. This compartment allows the generation of peptides that associate with MHC class II antigens and are recognized by CD4[+]-T cells[36,37]. This pathway, called exogenous pathway, does allow the generation of NP peptide from 91A3-NP antibody but, as seen below, it allows the generation from 91A3-HA of an HA peptide that associates with I-E^d MHC class II antigen and is recognized by an HA specific T helper hybridoma.

As mentioned above we generated a 91A3 antibody that carries in place of the heavy chain D segment a T helper epitope of the h agglutinin (HA) of PR8 influenza A virus. To test the ability of this antibody to mediate recognition by T cells we used a T hybridoma specific for site 1 (amino acid residues 110-120) of influenza virus HA. Once these cells recognize the HA peptide-I-E^d class II complex on the surface

Table 4. Lysis of transfectomas cells expressing 91A3-NP genes
by NP-specific CTL line.

target	% specific cytotoxicity
SP2/0	2.4 ± 2.2
SP2/0-91A3	1.9 ± 1.7
SP2/0 coated with NP peptide	43.0 ± 10.0
SP2/0-91A3-NP	30.0 ± 10.0

Target cells were coated with peptide where needed, labeled
with chromium and incubated with NP-specific CTL. The %
specific cytotoxicity is the mean ± SD of pooled data from 4
separate experiments.

of APC, they are activated and secrete IL3. The experiment we
performed to assess the ability of 91A3-HA to mediate
recognition by this T cells is based on measurement of IL3
secretion. For this purpose we used the IL3 dependent DA-1
cells. In the presence of IL3, DA-1 cells proliferate and are
able to cleave the tetrazolium ring of MTT dye which lead to
a purple precipitate that can be dissolved and measured by
colorimetry. As shown in Table 6 the 91A3-HA antibody
mediated T cell activation like HA-peptide and PR8 virus. This
activation is specific because 91A3 and 91A3-NP antibodies
which do not carry the HA epitope did not induce activation of
the T cell like NP peptide. These results suggest that the
91A3-HA antibody was taken up by spleen cells, processed
within the endosomal compartment and an HA peptide was
generated and then recognized by the HA-specific hybridoma.
In a separate experiment we found that the peptide generated
from 91A3-HA is recognized in association with I-E^d class II
antigen because an anti-I-E^d but not an anti-I-A^d inhibited the
activation of T cell mediated by either HA-peptide or 91A3-HA
antibody. We also found that the internalization of 91A3-HA
antibody into spleen cell is mainly mediated through Fcγ
receptor. These results indicate that 91A3-HA is an Ab2β
antibody that carries the HA peptide which can in turn be
generated from the Ig molecule, associate with MHC antigen and
induce activation of specific T cells.

Table 5. Soluble 91A3-NP antibody cannot mediate lysis of
SP2/0 cells by NP-specific CTL.

pulse	%specific cytotoxicity
nil	1.0
91A3	1.1
91A3-NP	1.1
NP peptide	47.2

SP2/0 cells were pulsed for 3 hours with either 1mg/ml of
antibodies or 5 ug/ml of NP peptide, washed, cultured for 24
hours, labeled with chromium and used as target for CTL.

Table 6. Activation of HA-specific T helper hybridoma by APC pulsed with 91A3-HA antibody

Antigen	Activation index
nil	1.0
91A3	1.2
NP-peptide	1.2
91A3-NP	1.1
HA-peptide	7.0
PR8 virus	7.6
91A3-HA	7.1

Balb/c spleen cells were incubated with either 71 ug/ml of peptide, 15 ug/ml of Ig or 2 ug/ml of UV-inactivated PR8 virus and HA-specific T helper hybridoma. After 2 days supernatant were collected and tested for IL3 content using the MTT assay and DA-1 cell line. The index of activation is the ratio of OD 570 obtained in presence of antigen over the OD570 obtained in absence of antigen.

CONCLUSION

The data presented in this review point out a new strategy to generate functional Ab2β type or internal image bearing antibodies. Indeed, we show that expression of T cell epitopes on Ig molecules is achievable via molecular immunology. These molecules can be defined as Ab2β antibody because they fulfill the strictest structural criterion and, as demonstrated, they imitate the function of the epitope as if these epitopes were expressed in viral proteins. So far this strategy is valid at least for the T cell epitopes used in these studies. It is possible that other T cell epitopes might require specific flanking regions in order to be generated. A definitive generalization of these strategies can not at this time be evaluated. It is also noteworthy that aside from maintaining a structure permissive to heavy and light chain pairing, there does not appear to be configurational requirements of the T cell epitope within the Ig molecule. Perturbation of heavy and light chain pairing would prevent Ig secretion. Thus, It is possible that this consideration may limit the length of the epitope to be inserted. On the other hand while the presence of linear T cell epitope within the Ig appears sufficient for recognition by T cell, it is possible that responses to a B cell epitope require a strict configurational conformity. The ability of B cells to recognize intact antigen leads us to assume that the epitope should be accessible in order to be "recognizable" to a B cell receptor.

REFERENCES

1. Jerne, N. K. (1976). Harvey Lectures Series 70. Academic Press, New York, San Francisco, London. P 93.

2. Kennedy, R. C., & Dressman, C. R. (1984). J. Exp. Med. 159: 655.
3. Bona, C., & Kohler, H. (1984). In Monoclonal and Anti-idiotypic antibodies: Probes for receptor structure and Function. Alan R. Liss. P 149.
4. Bona, C. A., Finley, S., Waters, S., & Kunkel, H. G. (1982). J. Exp. Med. 156: 986.
5. Hiernaux, J. (1988). Infection & Immunity.56: 1407.
6. Kennedy, R. C., Melnick, J. L., & Dressman, G. R. (1986). Scientific American. 225:48.
7. Idiotype Networks in Biology and Medicine, International Congress Series 862 (Oaterhaus, A.,& Uytdehaag, F. Eds). (1990). Publisher: Excerpta Medica Amsterdam. New York. Oxford. Rev. 90:106.
8. Anti-idiotypic Vaccines, Progress in Vaccinology Vol.3 (P. A. Cazenave, Ed). (1991) Publisher: Spring-Verlag, New Yor-Berlin-Heidelberg-London-Paris-Tokyo-Hong Kong-Barcelona.
9. Oudin, J., (1966). J. Cell. Phys. 67:77.
10. Bona, C. (1988). In biological Application of Anti-idiotypes. Vol I (Bona, C. Ed). Publisher: CRC Press.
11. Claflin, G. L., Lieberman, R., & Davies, J. M. (1974). J. Exp. Med. 139: 58.
12. Lancki, D. W., Lorber, M. I., Lochen, M. R., & Fitch, F. W.(1983). J. Exp. Med. 157: 921.
13. Lieberman, R., Potter, M., Humphrey, W. Jr., Muskinski, E. B., & Vrana, M. (1975). J. Exp. Med. 142: 106.
14. Waters, S. J., & Bona, C. (1988). Cell. Immunol. 111: 87.
15. Kimmura, H., & Wilson, D. B. (1984). Nature. 308: 463.
16. Ertl, H. C. J., & Bona, C. (1988). Vaccine. 6: 80.
17. Bona, C. A., Heber-Katz, E., & Paul, W. E. (1981). J. Exp. Med. 153: 951.
18. Zaghouani, H., Bonilla, F. A., Meek, K., & Bona, C. (1989). Proc. Natl. Acad. Sci. (USA). 86: 2341.
19. Bona, C. (1985). Clin. Immunol. Newsletter. 6: 87.
20. Regulatory Idiotopes, Modren Concept in Immunology Vol II (Bona, C. Ed). (1987). Publisher: John Wieley & Sons. New York-Chichester-Bribane-Toronto-Singapore.
21. Sege, K., & Peterson, P. A. (1978). Nature. 271: 167.
22. Biological Application of Anti-idiotypes, Vol I. (Bona, C. Ed) (1988). Publisher: CRC Press, Boca Raton, Florida.
23. Schiff, C., Millili, M., Hue, I., Rudikoff, S., & Fougereau, M. (1986). J. Exp. Med. 163: 573.
24. Bruck, C., Co, S. M., Slaoui, M., Gaulton, G. N., Smith, T., Fields, B. N., Mullins, J. I., & Green, M. I. (1986). Proc. Natl. Acad. Sci. 83: 6578.
25. Jerne, N.K., Roland, J., & Cazenave, P. A. (1982). EMBO. J. 1:243.
26. Palker, T. J., Clark, M. E., Langlois, A. J., Matthews, T. J., Weinhold, K. J., Rendall, R. R., Bolognesi, D. P., & Haynes, B. F. (1988). Proc. Natl. Acad. Sci. (USA). 85: 1932.
27. Zaghouani, H., Goldstein, D., Shah, H., Anderson , S., Lacroix, M., Dionne, G., Kennedy, R., & Bona, C. (1991). Proc. Natl. Acad. Sci. (USA). 88: 5645.
28. Zaghouani, H., Hall, B., Shah, H., & Bona, C. (1991). In Advances in Experimental Medicine and Biology Vol 303, Immunobiology of Proteins and Peptides VI. Human

Immunodeficiency Virus, Antibody Immunoconjugates, Bacterial Vaccines, and Immunomodulators. (Atassi, M. Z. Ed). Publisher: Plenum Press. New York and London.

29. Rathbun, G. A., Otani, F., Milner, E. C. B., Capra, J. D. & Tucker, P. H. W. (1988). J. Mol. Biol. 202: 383.

30. Sanz, I., & Capra, J. D. (1987). Proc. Natl. Acad. Sci. (USA). 84: 1085.

31. Taylor, P. M., Davey, J., Howland, K., Rothbard., J. B., & Askonas, B. A., (1987). Immunogenetics. 26: 267.

32. Haberman, A. M., Moller, C., McCreedy, D., & Gerhard, W. U. (1990). J. Immunol. 145: 3087.

33. Zaghouani, H., Krystal, M., Kusu, H., Moran, T., Shah, H., Kuzu. Y., Schulman, J., & Bona, C. (1992). J. Immunol. In press.

34. Gillies, S. D., Morrison, S. L., Oi, V. T., & Tonegawa, S. (1983). Cell. 33: 717.

35. Landolfi , N. F., Capra, J. D., Tucker, P. H. W. (1986). J. Immunol. 137: 362.

36. Unanue, E. R., Bellen, D. I., Lu, C.Y., & Allen P. M. (1984). J. Immunol. 132: 1.

37. Townsend, A. R. M., Rothbard, J., Gotch, F. M., Bahadur, G., Wraith, D., & McMichael, A. J. (1986). Cell. 44: 959.

14

CLONED SUPPRESSOR T CELLS --- THEIR IDENTITY, FUNCTIONS, AND MEDIATORS

Tomio Tada, Tadahiro Inoue, Shuichi Kubo, and Yoshihiro Asano

Department of Immunology, Faculty of Medicine University of Tokyo, Tokyo, Japan

Introduction

The existence of a specialized T cell subset suppressing the immune response is currently a controversial matter (1,2). There have been no specific cell surface markers determined on suppressor T cells (Ts) that distinguish Ts from helper T (Th) and cytotoxic T (Tc) cells. No particular lymphokines have been identified that mediate the effects of Ts. The general rule determining antigen- and MHC-restriction specificities of Ts has not been well established. The nature of the I-J molecule associated with various Ts functions has not been clarified. Nevertheless, the importance of suppression by T cells in the immune system is obvious because of the limitation of clonal deletion in the maintenance of self tolerance.

The difficulties underlying in defining Ts are primarily due to the paucity of reliable Ts clones. Only very few examples of Ts clones have been characterized (3-11). Ts cell hybridomas have been found to be extremely unstable (12). In order to define the nature and function of Ts as well as mediators produced by them, the establishment of stable Ts clones is definitely necessary. During the past few years we have tried to establish cloned T cells that have definite suppressor functions without helper and killer activities (4, 13-16). This paper describes some salient features of cloned Ts cells, and discusses the role of Ts in the regulation of the immune response.

Establishment of suppressor T cell clones

We would define suppressor T cells as *"T cells specialized in suppressing the immune responses --- not by the cytotoxicity for antigen-presenting cells (APC) or Th cells and not by the dose effect of known helper type interleukins"*. There are a few examples of suppression caused by the cytotoxicity of Tc cells which recognize antigen on APC or Th cells (17).

T Lymphocytes: Structure, Function, Choices, Edited by F. Celada and B. Pernis, Plenum Press, New York, 1992

However, such Tc cells may not be called as Ts in the strict sense. Interferon γ (IFNγ) and transforming growth factor β (TGFβ) have been claimed to be suppressive under limited condition (18,19), but they are not legitimate mediators of Ts cells, as such interleukins are produced by heterogeneous cell types other than Ts. Other interleukins such as IL4 is also suppressive at a high dose (20). IL10 also inhibits the synthesis of interleukin of Th1 cell type by acting on APC (21), but it has many other functions such as growth facilitation of mast cells. Thus, known interleukins may not be called as legitimate suppressor factors.

In order to characterize Ts cells at the clonal level, we have established a number of IL2-dependent T cell clones of different origins and different antigen-specificities (4, 13-16). These include both CD4+ and CD8+ T cell clones with helper and suppressor functions. CD4+ T cell clones were generally established by the in vitro stimulation of antigen-primed T cells with homologous antigen and APC followed by manifestation and cloning in IL2 containing medium (4,15). CD8+ T cell clones were made with a small modification (13): Antigen-primed splenic T cells were depleted of CD4+ T cells by cytotoxic treatment with anti-CD4 and C, and resulting CD8+ T cells were manifested by antigen plus APC followed by a culture in IL2-containing medium. They were cloned and maintained by periodical stimulation with antigen and with frequent changes of IL2-containing culture medium. Some Tc clones have been provided by Dr. N. Shinohara of the Mitsubishi-Kasei Institute of Life Science, Tokyo, Japan.

Table 1. Production of cytokines after stimulation with immobilized anti-CD3 by CD4+ and CD8+ T cell clones with different functions

Clone	Phenotype	Category	Cytokines						
			IL2	IL3	IL4	IL5	IL10	GMCSF	IFNγ
28-4	CD4	Th1	+	-	-	-	-	++	-
2-15-5	CD4	Th1	+	++	-	-	-	+	++
24-2	CD4	Th2	-	++	+	++	++	++	-
9-5	CD4	Ts	±	+	-	-	-	+	++
25-18	CD4	Ts	±	+	-	-	-	+	++
QM56	CD4	Tc	-	++	-	+	-	+	++
B02	CD4	Tc	-	++	-	-	-	+	++
HD-8	CD8	Ts	±	-	-	-	+	-	+
7C3	CD8	Ts	-	-	-	-	+	-	+
QM3	CD8	Tc	-	-	-	-	-	-	++

Production of cytokines were determined by enzyme-linked immunoassays or by Northern blotting. Categories were assigned by functional tests. No definite correlations were observed between their functions and the pattern of cytokine production.

The established T cell clones can be divided into several categories by functional tests: They are, CD4+ Th, CD4+ Ts, CD4+ Tc, CD8+ Ts and CD8+ Tc (Table 1). CD4+ Th clones are further divided into Th1 or Th2 subtypes by the pattern of lymphokine production: Th1 is characterized by their ability of producing IL2 and IFNγ, and Th2 by IL4 and IL5. There are several CD4+ T cell clones which have no helper activity in any of the experimental conditions but exert a definite suppressive activity in T cell-dependent antibody formation (4). They do not produce both IL2 and IL4, but IFNγ is produced at variable degrees. CD8+ T cell clones are generally Tc whose interleukin production pattern is variable. The fourth category is the CD8+ Ts. Ts clones belonging to this category show strong suppressor functions when added to the in vitro secondary antibody response or to antigen-induced IL2 production of antigen-primed spleen cells. They have no cytotoxic activity for Th clones, Con A blasts and APC. They are unable to kill L cells transfected with MHC genes that serve as the target molecules for specific recognition. They produce little IL2 and IL4, but IFNγ is produced upon stimulation with antigen and APC. CD8+ Ts clones invariably produce IL10, which is absent in any CD8+ Tc clones so far tested. Interestingly, all the CD8+ Ts clones established in independent experiments expressed a Vβ8 gene as a part of T cell antigen receptor (TcR) (13).

Properties of CD4+ Ts clones

Some of the established CD4+ T cell clones were unable to exert helper functions but exhibited a strong suppressive activity in the in vitro antibody response, and thus were defined as CD4+ Ts clones (4). When these clones were added to the *in vitro* secondary antibody response of syngeneic spleen cells, they showed a strong suppressive activity for the IgG antibody synthesis. However, if the clones were added to the response of allogeneic T and B cells, there were no suppression even in the presence of relevant APC and antigens. This indicates that the suppression induced by Ts clones has strict antigen- and MHC-restriction specificities (15).

We have attempted to study the mechanism of Ts-mediated suppression in an experimental system where intracellular Ca^{2+} ($[Ca^{2+}]i$) of interacting cells were directly measured by a stopped-flow fluorometory (22). In brief, antigen-specific MHC-restricted Th clones were labelled with a Ca^{2+}-sensitive fluorophore Fura 2 and admixed with antigen-pulsed syngeneic APC. This resulted in the fluorescence emission from the activated cells reflecting the increase of intracellular Ca^{2+}. Ts clones preactivated with antigen plus APC were added to this reaction mixture, and the changes in the fluorescence intensity of responding Th cells were measured by the stopped-flow fluorometory (22).

A rapid increase of $[Ca^{2+}]i$ in cloned Th cells was observed within a few seconds after the antigenic stimulation (Fig. 1). This Ca^{2+} response was dependent on the presence of antigen-pulsed MHC-matched APC. To see the effect of Ts on the increase of $[Ca^{2+}]i$ of Th clones, Ts clones were first stimulated with antigen-pulsed APC for different time lengths, and were then added to the Fura 2-loaded Th clones which was subsequently activated by the antigen-pulsed APC present in the mixture.

As shown in Fig. 1, the expected increase of intracellular Ca^{2+} in the Fura 2-loaded Th clone was strongly inhibited by the presence of an activated Ts clone. By combinations of Ts and Th clones with different antigen- and MHC- restriction specificities, we found that suppression was, in general, antigen-specific and MHC-restricted. In Fig. 1, we can see that a KLH plus I-A^k-specific Ts clone preactivated for 2 min with the antigen pulsed APC could suppress the Ca^{2+} influx of a Th clone from C3H that have the same antigen- and MHC-restriction specificities (closed squares). The Ts clone also could suppress the Ca^{2+} increase of KLH + I-A^b specific Th clone (closed circle) if it was stimulate with APC from (B6 x C3H)F$_1$, but it required more than 30 min for a maximal activation of the Ts clone. The same Ts clone preactivated for only 2 min showed less activity to suppress the response. If the Th clone was stimulated with B6 APC, the increase of $[Ca^{2+}]i$ of responding Th was substantially suppressed by the I-A^k-restricted Ts clone (open circles), but the maximum suppression was achieved only when Ts clones were preactivated for 180 min with antigen and APC. If both an antigen- and MHC-specificities were unmatched, no detectable suppression was observed (open squares).

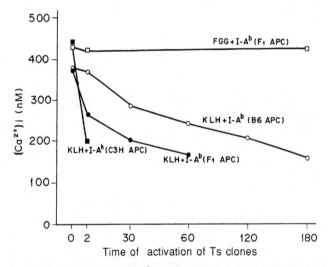

Fig.1. Suppression of Ca^{2+} influx of Th clones with different MHC- and antigen-specificities by a Ts clone preactivated for different period of time with antigen pulsed APC. The clones with indicated specificities were loaded with Ca^{2+} sensitive fluorophore Fura 2, and were mixed with a KLH and I-A^k-specific Ts cone which had been preincubated with KLH-pulsed APC for different times (horizontal axis). Note a strong suppression of Ca^{2+} influx of a Th clone having KLH plus I-A^k specificity by the Ts with the same specificity preactivated only for 2 min (■—■). The responses of a KLH plus I-A^b-specific Th clone stimulated by B6 (o—o) or (C3H x B6)F$_1$ APC (●—●) can be suppressed by the same Ts clone, but it requires a longer preincubation time to activate the Ts clone with KLH plus C3H APC to manifest the suppressive effect. If Th has different antigen (FGG)- and MHC (I-A^b)-specificities (□—□), no suppression was observed.

We also examined the inhibition of Ca^{2+} response by combinations of clones with different assigned functions. It was found that all Ts clones could suppress the Ca^{2+} response of antigen- and MHC-matched Th clones, while none of the Th clones had an ability to inhibit the Ca^{2+} responses of other T cells. Ts clones could suppress the response of Th clones, but they were unable to inhibit the response of other Ts clones. Thus, the suppression of Ca^{2+} response of Th by Ts was found to be *unidirectional* and *selective*.

These results indicated that CD4+ Ts clones suppress the specific activation of Th clones by inhibiting the early signal transduction involving the influx of Ca^{2+} ion into Th cells. Because of the longer time required for the suppression in MHC-unmatched combinations of Ts and Th clones, it was suspected that the effect should be mediated by a soluble factor released from Ts, which acts only on a specific receptor on Th cells. The factor should act only on target cells in close proximity, as the instant suppression occurs only when both Ts and Th have the same antigen- and MHC-restriction specificity.

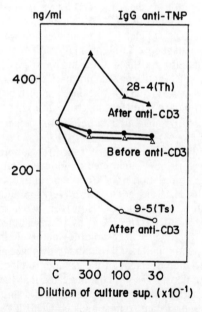

Fig. 2. Suppression of in vitro secondary antibody response by the culture supernatant of T cell clones obtained after stimulation with immobilized anti-CD3. Note that the culture supernatant of a CD4+ Ts clone 9-5 after the stimulation was able to inhibit the antibody response of primed spleen cells. The culture supernatant from a Th clone 28-4 obtained under the identical stimulation is enhancing the antibody response.

In fact, when CD4+ Ts clones were stimulated with immobilized anti-CD3, a strong suppressive activity was released into the culture supernatant (Fig. 2). The addition of a small quantity of the culture supernatant into the in vitro antibody response strongly reduced the amount of specific antibody produced by primed T and B cells *in vitro*. Such an activity was not detected in the culture supernatant of unstimulated CD4+ Ts. The appearance of the activity was within 6 hours after the anti-CD3 stimulation. None of the Th clones stimulated with anti-CD3 under the identical condition produced such a suppressive factor. The culture supernatant of Th clones after the anti-CD3 stimulation always enhanced the antibody response reflecting the release of IL2 or IL4 (Fig. 2). The factor suppressing the antibody response was different from IFNγ, TGFβ and IL10 as determined by absorption studies.

Properties of CD8+ Ts clones

The properties of CD8+ Ts clones are summarized as follows: CD8+ Ts clones showed a potent inhibitory activity on the antibody response induced B cells and normal or cloned Th cells (13). CD8+ clones were also able to inhibit the antigen-induced proliferative response of CD4+ Th clones. However, the proliferation of the same Th clones induced by Con A or IL2 was not inhibited. IL2 production of Th cells induced by antigenic stimulation was also inhibited by CD8+ Ts clones. The suppression was MHC-unrestricted and totally antigen-nonspecific. Interestingly, all of the CD8+ Ts clones established in our laboratory utilized Vβ8, and were class II-restricted. Both anti-class II and anti-CD8 antibodies could inhibit the response of Ts clones stimulated with relevant APC. CD8+ Ts clones can also be stimulated by anti-Vβ8 and anti-CD3 to induce proliferation. No cytotoxic activity against APC, Th clones, and class II-transfected L cells was detected before or after stimulation of CD8+ Ts (13).

Since CD8+ Ts clones exert a nonspecific suppression of Th cell-dependent responses, we have attempted to induce soluble mediators from CD8+ Ts clones by stimulation with Con A or antigen plus APC without success. These cells also did not constitutively produce any suppressive activity in the culture supernatant. However, when these clones were stimulated with insolubilized anti-CD3, they were able to release potent suppressive factors. They produced soluble factors within 4 - 8 hr after stimulation which potently inhibited the antibody response and antigen-induced proliferation of T cell clones. None of the Th and Tc clones in our collection produced such a suppressive factor. This indicates that the production of the suppressive factors by stimulation with anti-CD3 is a unique property of Ts clones. The culture supernatant from stimulated CD8+ Ts clones could inhibit the proliferative responses of Th clones when the latter were stimulated with antigen plus APC, or anti-TcRαβ monoclonal antibodies. The proliferative responses induced by Con A and IL2 were not inhibited by the supernatant from CD8+ Ts clones.

We have recently determined that one of the inhibitory factors released from CD8+ Ts clones is IL10 which inhibited the IL2 production of Th1 clones (Inoue et al., unpublished). None of the CD8+ cytotoxic T cell clones so far tested produced IL10. However, all of the suppressive activities were not ascribed to the function of IL10: CD8+ Ts clones could inhibit the function of both Th1 and Th2 clones, the latter of which is known to be refractory to the action of IL10. CD4+ Ts clones did not produce IL10 nor expressed IL10 mRNA. The culture supernatant was yet able to inhibit the in vitro antibody response in an antigen-nonspecific manner. IL10 had no suppressive effect on the in vitro secondary antibody response. The antibody against IL10 was unable to absorb the suppressive activities in the culture supernatant. Thus the inhibitory factors from both CD4+ and CD8+ Ts clones are different from IL10. None of the antibodies against IFNγ and TGFβ were able to block the suppressive activity of the culture supernatant of both CD4+ and CD8+ Ts clones.

Discussion

"Do specialized suppressor T cells exist?" is a long standing question waiting for an answer (1). It is now partially answered by the fact that specialized T cell clones with stable and definite suppressive functions have been established. These include both CD4+ and CD8+ Ts clones. They are distinct from Th1, Th2 and Tc, as they exert reproducible suppressor functions without producing helper type interleukins. The effect is not due to cytotoxicity for either APC or Th cells.

The suppression induced by Ts clones is due to the inhibition of early signal transduction involving the increase of intracellular Ca^{2+} which is initiated by the stimulation through TcR complex by antigen and APC (22). Some important features defining the Ts-mediated suppression have been defined: 1) Ts clones can suppress the Ca^{2+} response of Th clones of either Th1 and Th2 type, but not of other Ts clones (selectivity), 2) Th clones cannot suppress the Ca^{2+} response of Ts clones (unidirectionality), 3) suppression is more effective when antigen- and MHC-restriction specificities are identical between Ts and Th clones (proximity). It takes more time to make the Ts to inhibit the Ca^{2+} influx of Th clones if they are separately stimulated by APC having corresponding MHC-restriction specificities. 4) The final effect is mediated by antigen-nonspecific and MHC-unrestricted factor (non-specific suppression).

These figures of T cell mediated suppression cannot be explained by the cytotoxicity or by the effects of previously known interleukins including IFNγ and TGFβ. In fact, the suppressive effect of culture supernatant for in vitro antibody response cannot be abrogated by absorption with antibodies against known interleukins. The most feasible explanation is that Ts clones produce unique suppressor factors different from known interleukins that selectively act on Th cells expressing the receptor for suppressor factors. Our current effort is to characterize the factor and its receptor by molecular techniques.

However, the "suppressor factor" released from our Ts clones is different from antigen-specific suppressor T cell factor (TsF) described in previous studies. Antigen-specific factors are, in general, characterized by two properties, i.e., direct antigen-binding specificity and I-J determinants. It acts only on I-J compatible target cells (23, 24). The nonspecific factor which we described here has no antigen-binding activity and no I-J determinant. In our opinion, the established Ts clones of both CD4+ and CD8+ phenotypes are suppressor effector cells in the suppressor circuit. They are different from suppressor inducer cells producing antigen-specific factors. The cloning procedure for Ts cells used in the present study may have preferentially enriched the effector Ts.

However, the importance of I-J molecule in the suppressor circuit has been rediscovered in recent experiments with those T cell clones: We found that a monoclonal anti-I-J antibody could mimic the effect of Ts clones described here (25). If we pretreated Th clones with anti-I-J before stimulation with antigen and APC or anti-TcRαβ, both the proliferation and IL2 production of the Th clones were inhibited. The inhibition was due to the blockade in the early Ca^{2+} influx. Thus, I-J seems to be associated with the receptor for the suppressor factor, and anti-I-J should have acted as an agonist for the suppressor factor. I-J has been determined as a cell surface dimeric glycoprotein on Th clones (26), and is associated with a tyrosine kinase *fyn* (unpublished data). This tyrosine kinase is known to be associated with TcR/CD3 complex. I-J or I-J-associated receptor for the suppressor factor may act to inhibit the TcR-*fyn*-mediated signal transduction pathway by competing for unknown substrate of *fyn*.

The availability of specialized Ts clones of both CD4+ and CD8+ phenotypes will be a breakthrough for solving the long standing enigmas of suppression. The inhibition of signal transduction of only selected target cells by Ts and their products should open a possibility of selective immunosuppression of T cell-dependent immune responses having advertent effects. Only the responses induced by the recognition of specific antigens such as autoantigens through TcR can be suppressed by the factor produced by Ts clones. It can also be applied to produce a specific tolerance to certain antigens including alloantigen.

References

1. Möller, G. (1988) Do suppressor T cells exist? *Scand. J. Immunol.* **27**:247.
2. Tada, T., Asano, Y. and Sano, K. (1989) Present understanding of suppressor T cells. 26th Forum in Immunology. *Res. Immunol.* **140**:291.
3. Fresno, M., Nabel, G., McVay-Baudreau, L., Furthmayer, H. & Canter, H. (1980) Antigen-specific T lymphocyte clones. I. Characterization of a T lymphocyte clone expressing antigen-specific suppressive activity. *J. Exp. Med.* **153**:1946.
4. Asano, Y. & Hodes, R.J. (1983) T cell regulation of B cell activation. Cloned Lyt-1+2- Ts cells inhibit the major histocompatibility complex-restricted Th cell interaction with B cells and/or accessory cells. *J. Exp. Med.* **158**:1178.
5. Levich, J.D., Weigle, W.O. & Parks, D.E. (1984) Long term suppressor

cell lines. I. Demonstration of suppressive function. *Eur. J. Immunol.* **14**:1073.

6. Nakauchi, H., Ohno, I., Kim, M., Okumura, K. & Tada, T. (1984) Establishment and functional analysis of a cloned, antigen-specific suppressor effector T cell line. *J. Immunol.* **132**:88.

7. Heuer, J. & Kölsch, E. (1986) Selective elimination through a cytolytic mechanism of bovine serum albumin-specific T helper lymphocytes by T suppressor cells with the same antigen specificity. *Eur. J. Immunol.* **16**:400.

8. Modlin, R., Kato, H., Mehra, V., Nelson, E.E., Fan, X.D., Rea, T.H., Pattengle, P.K. & Bloom, B.R. (1986) Genetically restricted suppressor T-cell clones derived from lepromatous leprosy lesion *Nature* **322**:459.

9. Mohagheghpour, N., Damle, N.K., Takada, S. & Engleman, E.G. (1986) Generation of antigen-specific suppressor T cell clones in man. *J. Exp. Med.* **164**:950.

10. Hisatsune, T., Enomoto, A., Nishijima, K., Minai, Y., Asano, Y., Tada, T. & Kaminogawa, S. (1990) CD8+ suppressor T cell clone capable of inhibiting the antigen- and anti-T cell receptor-induced proliferation of Th clones without cytotoxic activity. *J. Immunol.* **145**:2421.

11. Takata, M., Maiti, P.K., Kubo, R.T., Chen, Y.H., Holford-Strevens, V., Rector, E.S. & Sehon, A.H. (1990) Cloned suppressor T cells derived from mice tolerized with conjugates of antigen and monomethoxypoly-ethylene glycol. Relationship between monoclonal T suppressor factor and the T cell receptor. *J. Immunol.* **145**:2846.

12 Taniguchi, M., Saito, T. & Tada, T. (1979) Antigen-specific suppressive factor produced by a transplantable I-J bearing T-cell hybridoma. *Nature* **278**:555.

13. Hu, F.Y., Asano, Y., Sano, K, Inoue, T., Furutani-Seiki, M. & Tada, T. (1992) Establishment of stable CD8+ suppressor T cell clones and the analysis of their suppressive function. *J. Immunol. Methods* (in press).

14 Nakayama, T., Kubo, R.T., Kubo, M., Fujisawa, I., Kishimoto, H., Asano, Y. & Tada, T (1988) Epitopes associated with MHC restriction site of T cells. IV. I-J epitopes on MHC-restricted cloned T cells. *Eur. J. Immunol.* **18**:761.

15. Asano, Y., Shigeta, M., Fathman, C.G., Singer, A. & Hodes, R.J. (1982) Role of the major histocompatibility complex in T cell activation of B cell subpopulations. A single monoclonal T helper cell population activates different B cell subpopulations by distinct pathways. *J. Exp. Med.* **156**:350.

16. Tada, T., Inoue, T. and Asano, Y. (1992) Suppression of the immune responses by cloned T cells and their products. *Behring Inst. Mitt.* **9**:78

17. Shinohara, N., Huang, Y.-Y. & Muroyama, A. (1991) Specific suppression of antibody responses by soluble protein-specific, class II-restricted cytolytic T lymphocyte clones. *Eur. J. Immunol.* **21**:23.

18. Mossman, T.R. & Coffman, R.L. (1989) Th1 and Th2 cells: Different patterns of lymphokine secretion lead to different functional properties. *Ann. Rev. Immunol.* **7**:145.

19. Wahl, S.M., Hunt, D.A., Wong, H.L., Dougherty, S., Mccartney-Francis, N., Wahl, L.M., Ellingsworth, L., Schmidt, J.A., Hall, G., Roberts, A.B. & Sporn, M.B. (1988) Transforming growth factor-β is a potent immunosuppressive agent that inhibits IL-1-dependent lymphocyte proliferation. *J. Immunol.* **140**:3026.
20. Asano, Y., Nakayama, T., Kubo, M., Fujisawa, I., Karasuyama, H., Singer, A., Hodes, R.J. & Tada, T. (1988) Analysis of two distinct B cell activation pathways mediated by a monoclonal T helper cell. II. T helper cell secretion of interleukin-4 selectively inhibits antigen-specific B cell activation by cognate, but not non-cognate, interactions with T cells. *J. Immunol.* **140**:419.
21. Fiorentino, D.F., Zlotnik, A., Vieira, P., Mosmann, T.R., Howard, M., Moore, K.W. & O'garra, A. (1991) IL-10 acts on the antigen-presenting cell to inhibit cytokine production by Th1 cells. *J. Immunol.* **146**:3444.
22. Utsunomiya, N., Nakanishi, M., Arata, Y., Kubo, M., Asano, Y. & Tada, T. (1989) Unidirectional inhibition of early signal transduction of helper T cells by cloned suppressor T cells. Int. Immunol. 1:460.
23. Dorf, M.E., & Benacerraf, B. (1985) I-J as a restriction element in the suppressor T cell system. *Immunol. Rev.* **83**:23.
24 Asherson, G.L., Colizzi, V. & Zembala, M. (1986) An overview of T-suppressor cell circuits. *Ann. Rev. Immunol.* **4**:37.
25. Asano, Y., Nakayama, T., Kishimoto, H., Komuro, T., Sano, K., Utsunomiya, N., Nakanishi, M. & Tada, T. (1989) Cell membrane molecule I-J transduces a negative signal for early T-cell activation induced via the TCR. Cold Spring Harbor Symp. Quant. Biol. 54:683.
26. Nakayama, T., Kubo, R.T., Kishimoto, H., Asano, Y. & Tada, T. (1989) Biochemical identification of I-J as a novel dimeric surface molecule on mouse helper and suppressor T cell clones. *Int. Immunol.* **1**:50.

15

THE AUTOREACTIVE T CELL RECEPTOR:

STRUCTURE AND BIOLOGICAL ACTIVITY

Ellen Heber-Katz

The Wistar Institute
Philadelphia, PA 19104

Introduction

Autoimmune disease animal models fall into two categories. In the first case, there is spontaneously occurring autoimmune disease (such as diabetes). In this case, animals have been identified which are particularly susceptible to the disease (for diabetes there is the NOD mouse and the BB rat). In the second case, there is the experimentally-induced autoimmune disease. This form of disease model makes use of an identified self-antigen which will cause disease symptoms upon injection into susceptible animals. Furthermore, the immune and antigen-specific components involved in the disease can be identified.

We have chosen to study the latter case and specifically the experimentally induced disease, experimental allergic encephalomyelitis or EAE (a model for multiple sclerosis) which is induced by injection of the myelin-associated CNS antigen, myelin basic protein or MBP.

I. The T cell repertoire in EAE

EAE is in fact a T cell-mediated disease. Thus, antigen-specific T cells can be isolated from MBP-primed animals and the disease can be adoptively transferred to naive animals with activated T cells alone. The T cell response to autoantigen follows the same rules as the response to exogenous antigen.
Namely, the response is both MHC-restricted and specific for an immunodominant antigenic determinant. Immunization with a whole protein in adjuvant leads to a T cell response to only one or a few antigenic determinants. Thus, the Lewis rat (the animal used in our studies) when immunized with whole myelin basic protein in CFA responds predominantly (approximately 75% of the primary MBP-specific T cell response) to an antigenic determinant found in residues 68-88 of MBP. This determinant is also the dominant encephalitogenic determinant, though

T Lymphocytes: Structure, Function, Choices, Edited by F. Celada
Plenum Pernis, Plenum Press, New York, 1992

other determinants have been defined (Offner et al., 1989) and the cells that transfer disease are class II restricted CD4+ T cells (Happ and Heber-Katz, 1987).

Our initial experiments examined one 68-88-specific Lewis rat T cell hybridoma. Sequencing of T cell receptor mRNA revealed a set of V regions similar to the Vβ8.2 and Vα2 families identified in the mouse. Hybridization of other 68-88 specific T cells showed that all of these cells used Vβ8 and most used Vα2 (Burns et al, 1989). We then sequenced the TcR mRNA from a large panel of 68-88-specific Lewis rat T cells and found that in fact not only was the identical Vβ8 used but also the junctional (CDR3) regions of the beta chains were very similar. The J regions themselves did not seem to be selected (Zhang and Heber-Katz, 1992).

It is known in the mouse and human that the Vβ8 family consists of three family members. To determine if the Vβ8 used by the MBP 68-88 T cells might be the only family member in the rat, we sequenced mRNA from various normal T cell populations in the rat using a Vβ8 primer. We identified four Vβ8 family members, Vβ8A-D. The Vβ8.2 or Vβ8A (that used by the MBP 68-88 specific T cells) subfamily represented approximately 10% of the Vβ8 found in the cervical lymph nodes whereas approximately 100% of encephalitogenic T cells. One Vβ8 family member is particularly interesting in that it lacks a cysteine at residue 90. This cysteine is the second half of a disulfide intrachain bridge (not inolved in the alpha-beta interchain disulfide bridge) which may be involved in TcR loop formation and proper folding of the TcR. The cysteine has been replaced by a tryptophan in every case which is then followed by a proline (Zhang and Heber-Katz, 1992). This Vβ8 which we call Vβ8C has been identified from lymphoid tissue from neonates and adults as well as from nude rats (Table 1). In all of 6 cases we have identified thus far, the sequence is in-frame supporting the possibility that it is functional and is used. We have however not identified a T cell using this sequence so that we can determine its fine specificity or analyze the alpha chain used. It would not be surprising though that this is a part of a functional receptor since immunoglobulin chains have been identified which lack a cysteine at the same position (Rudikoff and Pumphrey, 1986).

II. The V Region Disease Hypothesis

The V region combination used by MBP-reactive encephalitogenic T cells, Vα2Vβ8, is interestingly used not only by Lewis rats, but it has also been shown to be used B10.Pl mice and 4 other rat strains as well (Heber-Katz and Acha-Orbea, 1989). This is true even though the antigenic determinants of MBP used by these other animals are different from residues MBP 68-88 and the MHC-restricting elements are different.

This unusual finding led us to propose that this V region combination may be involved in the binding to another ligand, other than the MHC-antigenic peptide complex and that this ligand may act as a homing receptor or a molecule involved in pathogenicity and might be found in CNS tissue

Table 1.   COMPARISON OF THE Vβ8A and Vβ8C
NUCLEOTIDE AND AMINO ACID SEQUENCES

	V region	nDn region	J region
I. Vβ8A			
Adult	TGT GCC AGC AGT		
	C A S S		
II. Vβ8C			
Neonatal C	TG(T)G CCA GCA GT	G ACA	TAT (Jβ 2.1)
	W P A		
Adult C	TG(T)G CCA GC	C GGA CAG GGA GGG TG	C AAC (Jβ 1.5)
	W P A		
	TG(T)G CCA GCA G	GA CTG GGG GGG CGC	AAA (Jβ 2.2) *
	W P A		
Nude C	TG(T)G CCA G	GC ACC GAA TGG GGG	AAC (Jβ 1.1) *
	W P G		

From S. Goldman and X-M Zhang, preliminary data

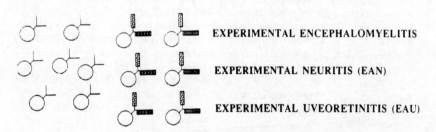

EXPERIMENTAL ENCEPHALOMYELITIS

EXPERIMENTAL NEURITIS (EAN)

EXPERIMENTAL UVEORETINITIS (EAU)

Figure 1. Sharing of TcR Genes (Vα2Vβ8) by Autoimmune T cells

(Heber-Katz and Acha-Orbea, 1989, Heber-Katz, 1990, and Kawai et al.,1990, Lider et al. 1991).

However,Vα2Vβ8 has also been identified in T cell populations and clones from Lewis rats specific for other pathogenic antigens in other disease models (Figure 1). The T cells specific for these antigens have been shown to cause experimental allergic uveoretinitis (the S-antigen, a retinal protein)(Gregerson et al.,1991, Merryman et al. 1991) and experimental allergic neuritis (the P2 protein from the peripheral nerve myelin sheath) (Desquenne-Clark, Heber-Katz, Rostami, 1992). Furthermore, the T cells involved in adjuvant arthritis also appear to use similar V regions (Clark, unpublished data). The usage of the same V region combination has led us to examine both nervous system-related tissue and autoimmune target tissue in general for the putative second ligand.

III. The Regulation of the Immune Response

EAE is an acute disease which occurs after the injection of encephalitogenic T cells or encephalitogenic antigen into mice or rats. The disease symptoms rapidly appear after injection of cells (3-5 days) or antigen (11-13 days) and then just as rapidly subside. This rapid recovery has been attributed to something other than the disappearance of antigen and there are several reasons for believing that not only rapid recovery but also occurrence and more specifically recurrence of disease is due to active regulation of the relevant immune responses. Thus, it was found that after a bout of EAE, animals become resistant to rechallenge with the encephalitogen, MBP. The phenomenon of T cell vaccination, explored by Cohen and colleagues (Ben-Nun et al , 1981), is based on this finding and it was shown that T cells which are unable to cause EAE because they are attenuated by x-irradiation or mitomycin or pressure treatment will cause the same resistance to rechallenge as seen in active disease. It was proposed that T cell vaccination worked because the animal became immunized against an antigen on the T cell, this antigen being most likely the T cell receptor.

Subsequent studies as described above have shown that restricted T cell usage allows for potential TcR V region-specific therapy. Anti-V region or anti-idiotypic antibody has been shown to modulate EAE (Owhashi and Heber-Katz, 1988). An interesting approach has been the use of peptides derived from EAE-specific TcRs. In the original experiments the pre-injection of TcR peptides resulted in suppression of disease and the induction of regulatory T cells specific for those peptides and the TcRs from which they were derived (Howell et al, 1989, Vandenbark et al., 1989). The implication is that T cells present their receptors as processed peptides and that other T cells can recognize them on the surface of T cells and thus regulate the T cell response this way.

Our laboratory used the same TcR peptides described in the previous studies and derived from the Lewis rat TcR (Burns et al, 1989) alpha chain CDR3 region and the beta chain CDR2 region to modulate EAE. We found that the animals after having received TcR peptide in CFA and then challenged with MBP one month later displayed enhanced and chronic EAE, not suppression (Desquenne-Clark et al., 1991) This result at first appears contradictory but in fact could be interpreted as providing support for a TcR-specific regulatory cell. Thus, one could imagine that such a regulatory cell in our experiments are being anergized so that upon induction of EAE there is less regulation and more EAE (Figure 2). We have not been able to isolate cells from animals immunized with TcR peptide that can respond to that peptide.

In an attempt to test this, a conventional method of tolerance induction was used. Thus, we injected TcR peptide in saline into Lewis rat 24 hour neonates hoping to tolerize cells reactive to the peptide. These animals were then injected as adults with MBP in CFA and followed for symptoms of EAE. We found that those animals that received either saline alone or a control TcR peptide from a nonvariable framework region of the beta chain displayed the normal course of disease. However, if the animals received the beta chain CDR2 region TcR peptide, disease appeared earlier and the animals did not completely recover but rather cycled through disease symptoms (Figure 3). Some of the animals displayed symptoms of adjuvant arthritis supporting the notion that the same V regions are used and those T cells have become de-regulated. The fact that the results of TcR peptide priming in the adult look very similar to what is seen when rats are injected as neonates with TcR peptide support the idea that regulatory cells are being tolerized in some manner. Further support for this is found in the fact that in both cases, we have been unable to isolate T cells which can proliferate in-vitro to TcR peptide after immunization with TcR peptide in CFA.

This experiment has been repeated using the junctional region CDR3 TcR peptide from the alpha chain as the tolerogen. In this experiment, the control animals (they received saline or the control TcR peptide as neonates) as adults after injection of MBP in CFA displayed only mild EAE. The animals which received the CDR3 peptide and then MBP however displayed severe symptoms, 20% of them dying . Again there is evidence for adjuvant arthritis indicating that the alpha chain junctional CDR3 region used by MBP-reactive T cells is involved in this disease.

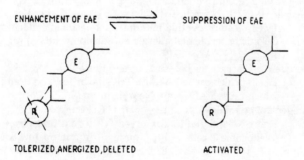

Figure 2. Regulation of EAE: A Double-edged Sword.
E: The effector or encephalitogenic T cell with the $V\beta 8 V\alpha 2$ TcR. R: The regulatory T cell which is specific for TcR peptides derived from $V\beta 8$ or $V\alpha 2$.

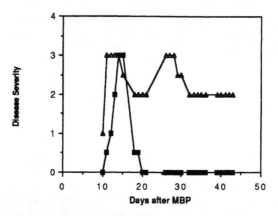

Figure 3. Enhancement of EAE.
Lewis rats were injected i.v. at birth with either saline (■) or the CDR2 beta chain- derived TcR peptide in saline (△). At three months of age, the animals were injected in the footpad with MBP in CFA and followed for symptoms of EAE.

The data thus far does support a "regulatory" cell which can be easily be tolerized. However, we have not yet been able to identify this regulatory T cell.

REFERENCES

Ben-Nun, A., H. Wekerle, and I.R. Cohen. 1981. Vaccination against autoimmune encephalomyelitis with T-lymphocyte line cells reactive against myelin basic protein. Nature 292:60.

Burns, F.R., Li, X., Shen, N., Offner, H., Chou, Y.K., Vandenbark, A.A., and Heber-Katz, E. 1989. Both rat and mouse T cell receptors specific for encephalitogenic determinant of myelin basic protein use similar V α and V β chain genes even though the major histocompatibility complex and encephalitogenic determinants being recognized are different. J. Exp. Med. 169: 27-39.

Desquenne-Clark, L., Esch, T.R., Otvos, L., Jr., and Heber-Katz, E. 1991. T cell receptor peptide immunization leads to enhanced and chronic experimental allergic encephalomyelitis. Proc. Natl. Acad. Sci. USA 88: 7719.

Desquenne-Clark, L., Heber-Katz, E., and Rostami, A.M. 1992. Lew Rat T cells which cause EAN use TcR similar to those used by T cells which cause EAE. Ann. Neurol., in press.

Gregerson, D.S., Fling, S.P., Merryman, C.F., Zhang, X., Li, X., and Heber-Katz, E. 1991. Conserved TcR V gene usage by uveitogenic T cells. Clin. Immunol. Immunopath. 58: 154.

Happ, M.P. and Heber-Katz, E. 1987. Differences in the repertoire of the Lewis rat T cell response to self and non-self myelin basic proteins. J. Exp. Med. 167: 502-513.

Happ, M.P., Kiraly, A., Offner, H., Vandenbark, A., and Heber-Katz, E. 1988. The autoreactive T cell population in experimental allergic encephalomyelitis: T cell receptor β chain rearrangements. J. Neuroimmunol. 19: 191-204.

Happ, M.P., Wettstein, P.J., Dietzschold, B., and Heber-Katz, E. 1988. Genetic control of the development of EAE in rats. Separation of MHC and non-MHC gene effects. J. Immunol. 141: 1489-1494.

Heber-Katz, E. and Acha-Orbea., H. 1989. The V-region disease hypothesis: Evidence from autoimmune encephalomyelitis. Immunology Today 10: 164-169.

Heber-Katz, E. 1990. The autoimmune T cell receptor: Epitopes, idiotopes, and malatopes. Clin. Immunol. Immunopathol. 55: 1.

Howell, M.D.,Winters,S.T.,Olee,T.,Powell,H.C.,Carlo,D.J.,and Brostoff, 1989. Vaccination against experimental allergic encephalomyelitis with T cell receptor peptides. Science 246:668.

Kawai, K., Heber-Katz, E., and Zweiman, B. 1991. Cytoxic effects of myelin-basic protein reactive T cell hybridomas on oligodendrocytes. J. Neuroimmunol. 32: 75.

Lider, O., Miller, A., Miron, S., Hershkoviz, R., Weiner, H.L., Zhang, X.-M., and Heber-Katz, E. 1991. Non-encephalitogenic CD4 - CD8 - Vα2Vβ8.2+ anti-myelin basic protein rat T lymphocytes inhibit disease induction. J. Immunol. 147: 1208.

Merryman, C.F., Donoso, L.A., Zhang, X.M., Heber-Katz, E., and Gregerson, D. 1991. Characterization of a new, potent, immunopathogenic epitope in S-antigen which elicits T cells expressing Vβ8 & Vα2 genes. J. Immunol. 146: 75.

Offner, H., Hashim, G.A., Celnick, B., Galang, A., Li, X., Burns, F.R., Shen, N., Heber-Katz, E., and Vandenbark, A.A. 1989. T cell determinants of myelin basic protein include a unique encephalitogenic I-E restricted epitope for Lewis rats. J. Exp. Med. 170: 355.

Owhashi, M. and Heber-Katz, E. 1988. Protection from experimental allergic encephalomyelitis conferred by a monoclonal antibody directed against a shared idiotype on rat T cell receptors specific for myelin basic protein. J. Exp. Med. 168: 2153-2164.

Rudikoff, S. and J G. Pumphrey. 1986. Functional antibody lacking a variable- region disulfide bridge. PNAS 83:7875.

Vandenbark,AA,Hashim,G., and Offner,H. 1989. Immunization with a synthetic T cell receptor V-region peptide protects against experimental autoimmune encephalomyelitis. Nature 341:541.

Zhang, X-M. and Heber-Katz, E. 1992. Autoreactive T cells in adult Lewis rats appear to be products of early ontogeny. J. Immunol. 148: 746.

16

THE RELATIONSHIP BETWEEN DIABETES AND LYMPHOPENIA IN THE BB RAT

Sarah Joseph, Geoffrey Butcher, William Smith[*], and Joyce Baird[*]

Dept. Immunology I.A.P.G.R., Babraham
Cambridge, CB2 4AT, England, U.K.

[*]Metabolic Unit, Western General Hospital
Crewe Rd., Edinburgh, Scotland, U.K.

INTRODUCTION

The BB rat is an animal model of type 1 insulin dependent diabetes mellitus which is thought to have emerged after a spontaneous mutation in a non-inbred colony of Wistar rats at the Biobreeding laboratories in Ottawa in 1974.(ref 1 and 2) As a result of this a colony of diabetes prone (BBDP) animals was established and rats were sent around the world where similar colonies were founded. Some laboratories (in Worcester and Edinburgh for example) have established diabetes resistant sublines (BBDR) with a 0% incidence out of the same stock and maintained them alongside the BBDP colony to serve as a source of related unaffected control animals.

The incidence of disease in the various BBDP colonies is approx 70% (although there is one colony with a reported incidence of 100% at Organon, Os, The Netherlands). The onset of IDDM usually starts at around 90-100 days of age and there is no sex bias. The disease in the BB rat proceeds extremely rapidly leading to insulin dependence within approximately 7-14 days. Where necessary diabetic BBDP rats are maintained on exogenous insulin and can be managed satisfactorily with daily injections.

THE DISEASE IS AUTOIMMUNE

Evidence from numerous laboratories has confirmed that the disease is autoimmune in causation and that the targets of destruction are specifically the pancreatic ß cells within the islets of Langerhans since virtually all the other cell types within the pancreas are undamaged in the early phase of the disease process. The onset of diabetic symptoms is preceded by a lymphocytic infiltration into the pancreas which is clearly visible in biopsies from prediabetic animals. By the time that the first symptoms of diabetes appear however, there is little or no "inflammation" within the pancreas and only residual ß cells are left. A well documented series of events precedes the onset of symptoms and it is clear that lymphocytes are instrumental in the destruction of the insulin producing ß cells. Walker et al (3) have shown that the first cells to infiltrate the islets are ED1+ macrophages which are not normally resident in the pancreas. They are followed by CD4+ lymphocytes and then later on CD8+ cells and other lymphoid cells are seen. The ß cells are destroyed by an as yet undefined mechanism and there have been numerous plausible suggestions for candidate effectors as well as the proposal that "non specific" cytokines themselves do the deed, perhaps on account of some idiosyncratic sensitivity of ß cells to particular lymphokine combinations. (4). It seems that the infiltration subsides rapidly when the ß cells have been destroyed and this is when the rat requires exogenous insulin for survival.

It is possible to transfer diabetes from an animal that is insulin dependent to one which has not yet developed disease by transferring the CD4+ lymphocytes alone, (although the cells do have to activated by a mitogen first (5)) and this provides formal proof that autoreactive T cells specific for a ß cell antigen are instrumental in the destructive process. This is further supported by the fact that typical immunosuppressive procedures such as neonatal thymectomy

(6) and administration of cyclosporin A to BBDP rats (7) are well known to ameliorate the disease symptoms.

BBDP RATS ARE LYMPHOPENIC

A striking feature of virtually all BBDP rats thus far described is that they display profound lymphopenia (8-11). BBDP rats from colonies with a high incidence of disease are homozygous for a recessive gene which results in the virtual absence of mature peripheral T cells. The lymphopenia is characterised by a complete absence of CD8+ T cells and the absence of CD4+ T cells expressing the rat alloantigen RT6 (12). RT6 is a molecule of unknown function which appears on the surface of most T cells some time after they come out of the thymus and disappears upon activation (13). Until now, high incidence diabetes has not been seen without congenital lymphopenia and some workers have postulated that it is the absence of a putative RT6+ "regulatory cell" in lymphopenic BBDP rats which leads to the disregulation of the autoreactive CD4+ T cells which initiate the disease process. This hypothesis is supported by the remarkable findings of Greiner and his colleagues (14); They found that if the RT6+ cells were artificialy depleted (at 30 days of age) in BBDR rats (which would otherwise never develop diabetes) using a mAb against RT6, then it was possible to induce IDDM in these animals. It seems that the disease process in these RT6 treated rats is not identical to that seen in BBBP rats as, for example it is not sensitive to the protective effects of silica administration as is the spontaneous disease (15). It is pertinent to note here that all BBDR rats thus far analysed do not display this severe lymphopenia and so they differ from the BBDP rats in this respect.

GENETIC REQUIREMENTS FOR DIABETES IN BB RATS

Because high incidence diabetes has never been seen in a BBDP colony of rats in the absence of the lymphopenia gene it has been assumed that homozygosity at this locus is necessary for IDDM in this model. By extensive "breeding studies" Eleanor Colle and her colleagues have demonstrated that diabetes in the BB rat is also strongly associated with the RT1u alleles of the MHC class II region and this has been confirmed by other groups (17-19). They have shown that in certain strain combinations (RT1u X RT1x) F1 animals (U X X) are susceptible to disease on the BB background. They have also shown that a u haplotype derived from strains other than BB will confer susceptibility and that it is therefore unlikely that the BB rat carries a novel u class II molecule (20,21). This group have also described a third susceptibility locus which they have termed Pli for (Pancreatic lymphocytic infiltration) (22-24). In their crosses this gene behaved in a dominant fashion and controlled the degree of infiltration of lymphocytes both around and within the islets of Langerhans (insulitis) as well as around the pancreatic ductules. The latter appears as discrete foci within acinar tissue and can be seen without diabetes in the absence of other predisposing genes. They postulate that there is a requirement for this gene in association with the other two for diabetes in their colony of BB rats. Pli phenotypes in other BBDP colonies has not been documented.

THE EDINBURGH COLONY

The Edinburgh colony of BB rats was derived in 1982 from a nucleus of 6 rats suppied by P. Thibert from Ottawa. A breeding programme was established which resulted in the establishment of 2 lines of BB/Ed rats; the so called "mainline" (BBDP/Ed) which now has a stable incidence of 70% diabetes and the subline (BBDR/Ed) with 0% diabetes. The disease in the mainline animals has been well characterised and is consistent with that seen in other BBDP colonies.

AIM OF THIS STUDY

The aim of this study was to type the two lines of BB rats from Edinburgh as thoroughly as possible for the two well characterised susceptibilty loci for IDDM in this model. The animals were first typed at the MHC using a wide panel of monoclonal antibodies to known rat haplotypes (24). The degree of lymphopenia was assessed by extensive analysis of the lymphoid subsets present in the periphery of both lines and comparing them to control non BB rats.

154

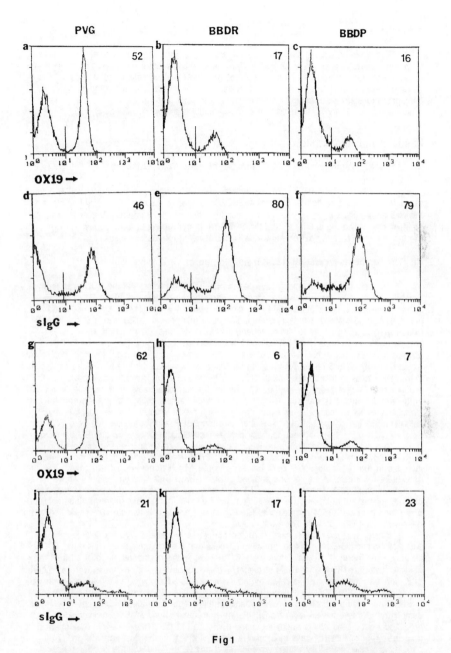

Fig1

Fig 1 (a)-(f) show lymph node lymphocytes (as prepared in materials and methods) from PVG, BBDR and BBDP rats stained with a T cell specific mAb (anti-CD5) in (a)-(c) and a B cell specific cocktail of mAbs (anti-sIgG) in (d)-(f). Fig 1 (g)-(l) shows peripheral blood leucocytes from PVG, BBDR and BBDP rats stained with the same mAbs. (g)-(i) shows the staining with the T cell specific mAb and (j)-(l) the staining with the B cell specific cocktail. The numbers in the top right hand corner of each box indicate the percentage of cells staining with the designated mAb. The marker is positioned on the basis of staining of the second stage reagent alone.

MATERIALS AND METHODS

The analysis was routinely performed using flow cytometry. Leucocytes were purified from either peripheral blood or cervical lymph nodes using routine procedures, the leucocyte fraction was purified using Ficoll hypaque (25) and the resultant cells were stained with a range of mAbs against rat cell surface molecules. The mAbs were detected using antisera conjugated to either fluorescein or phycoerythrin. See table 1 for mabs used in the subset analysis and ref 24 for mAbs used in the haplotyping experiment.

RESULTS AND DISCUSSION

(i) The MHC typing of the BBDR and BBDP rats from Edinburgh

The results of the haplotyping experiment (data not shown) revealed that, using a wide range of haplotype specific mAbs, both the BBDR and BBDP rats carry the RTI$^{\underline{u}}$ haplotype at the class I and class II loci. In this respect therefore they are the same as all other BB rats so far described . The result implied that the BBDR rat is not protected from disease by virtue of the fact that it is carrying a "resistant" non U haplotype at the class II locus.

(ii) The diabetes resistant rats from Edinburgh are lymphopenic

The subset analysis quickly revealed the unexpected finding that the BBDP rats from Edinburgh displayed the severe lymphopenia characteristic of diabetes prone BB rats. **Fig 1** shows some typical one dimensional FACS plots of leucocytes from cervial lymph node and peripheral blood stained with a T cell specific mAb (a)-(c) and (g)-(i) and a B cell specific mAb (d)-(f) and (j)-(l). (see figure legend for key). Panels (a)-(f) show the results obtained with lymph node lymphocytes and it was clear that both the BBDP and the BBDR animals from Edinburgh had a reduced proportion of T cells in their lymph nodes. A control rat would normally have 50-60% T cells in its lymph nodes whereas the two BB/Ed lines had less than 20% T cells in this compartment. This was mirrored by a corresponding increase in the proportion of B cells present , so that whereas in the control animals there were 40% B cells, in the two BB lines this was increased to 80%. Virtually all the cells in the lymph nodes were identifiably either B cells or T cells and the "gap" left by the missing T cells was filled by B cells. The striking finding however, was that both the BBDP and the BBDR rats from Edinburgh were displaying the lymphopenia normally only associated with diabetes prone BB animals.

The cellular composition of peripheral blood was more complicated in that although the proportion of T cells was greatly reduced in both BB/Ed lines, there was not a concomitant increase in the proportion of B cells (as seen in the lymph nodes). Fig (j)-(l) reveal that the percentage of B cells in the peripheral blood of both BB/Ed lines was only slightly above control values at 30%. The percentage of T cells however was very low at 5% compared with 60-65% for control animals. The "space "left by the "missing" T cells in the peripheral blood of the lymphopenic BB/Ed animals was filled by non-T non-B cells which were confirmed to be negative for the CD5 and sIg, (data not shown)

Similar analysis (not shown) revealed that there were virtually no CD8+ T cells in either of the two BB/Ed lines and that all the T cells that they do have were CD4+ and TCR2+, i.e they were "normal" T cells. There were CD8+ cells in the peripheral blood of all the animals tested and further analysis of these revealed that in control rats these were a mixture of T cells and NK cells (as indicated by the staining with 3.2.3), while in the lymphopenic rats from Edinburgh, the CD8+ cells were all NK cells. It is interesting to note that the CD8+ cells in BBDP rats play a significant role in diabetes in this model. Like and his colleagues demonstrated that adult pre-diabetic BBDP rats from their colony in Worcester could be protected from disease by antibody-mediated depletion of their CD8+ T cells . They inferred from their results that NK cells must play a significant part in the destruction of the pancreatic b cells since they were unable to detect any CD8+ T cells in the blood in these animals. In all respects the lymphopenia we have observed for the two Edinvurgh sublines is compatible with published descriptions.

(iii) The cellular composition of BB/Ed peripheral blood

Figs 2 and 3 show further flow cytometric analysis of the non-T non-B cells in the peripheral blood of the two BB rat lines from Edinburgh. Fig 2 (a)-(c) confirmed that the peripheral blood of the two lymphopenic BB/Ed lines contained approximately the same

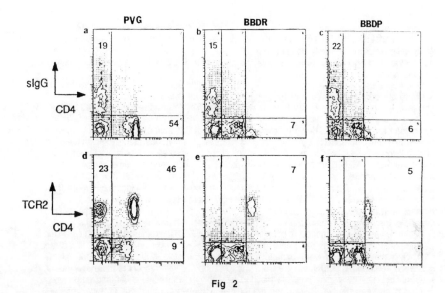

Fig 2

Fig 2 (a)-(c) show peripheral blood leucocytes from PVG, BBDR and BBDP rats stained with a B cell specific cocktail of mAbs in one dimension (RG7/9.1+ MARM4) and a CD4 specific mAb in the other dimension (OX38). Fig 2 (d)-(f) show the same cells stained with a TCR2 specific mAb against the same CD4 specific mAb to illustrate which of the CD4+ cells are T cells. The percentage of each population is indicated by the number in the boxes, and the markers are positioned in order to delineate the populations as clearly as possible.

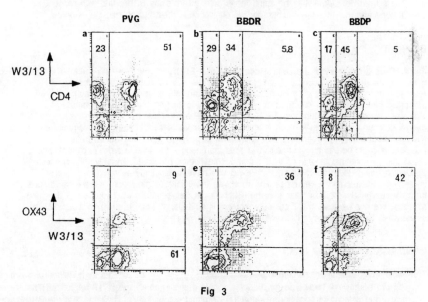

Fig 3

Fig 3 (a)-(b) show peripheral blood leucocytes from PVG, BBDR and BBDP rats stained with W3/13 (which is known to stain monocytes and T cells) and a mAb against CD4 (OX38). Fig 3 (d)-(f) show the same cells stained with OX43 (a monocyte specific mAb) and W3/13 (this time along the abscissa).The percentage of each population is indicated by the number in the boxes, and the markers are positioned in order to delineate the populations as clearly as possible.

proportion of B cells as blood from non lymphopenic controls. What was clear from these traces was that both the BBDR and the BBDP rats had a relatively expanded proportion of CD4 (dull) cells which were barely detectable in controls. These cells did not carry T cell markers as illustrated in fig 2 (d)-(f) which shows the proportion of CD4+ cells which are T cells in the rats analysed.

Fig 3 shows further characterisation of these CD4(dull) cells which represented such a significant proportion of cells in the peripheral blood of the BB/Ed rats. Fig 3 (a)-(c) shows that the cells stained very brightly with W3/13. This mAb, specific for CD43, is known to stain T cells and some monocytes as well as neutrophils. The W3/13+ CD4(dull) cells were larger than lymphocytes (as determined by their degree of FSC, data not shown). As shown in Fig3 (d)-(f) they did specifically stain with OX43, a mAb which was made after immunising a mouse with peritoneal cells from a rat and which recognizes a cell surface molecule of molecular weight 90,000 which is present on peritoneal macrophages, some alveolar macrophages and all vascular endothelium except that of brain capillaries (32). Mason et al found that 22% of PBL from normal rats stained with OX43, which is signifcantly more than the 8-10% detected in normal rats in these analysis. There may be a genetic variation between rat strains in the proportions of different monocyte subpopulations present and it may also depend of the immune status of the animal being examined.

It seemed that the monocyte population in both the BBDR and the BBDP blood consisted of several types of cells, as determined by their staining with OX41, OX42, OX43, W3/13 and CD4. Over the course of several experiments it was not possible to identify a reproducible difference between the BBDR and the BBDP peripheral blood with respect to the monocyte subpopulations present. It is possible that further analysis will reveal a genuine phenotypic difference between the two lines in the monocyte/macrophage populations in their peripheral circulation and it is also possible that such a phenotypic difference may have a physiological significance. Recently Kolb et al reported that the macrophages from their BBDP rats overproduced TNFα in response to LPS stimulation in vitro (36) These workers found that the peritoneal macrophages from their BBDP rats produced 10 X more TNFα than either control or BBDR macrophages when stimulated in vitro with LPS. The trait behave in a recessive manner in F1 crosses and so appeared to be regulated bya recessive gene. It is conceivable that the BBDP rats from Edinburgh may be significantly different from the BBDR rats with respect to such a gene and that this may account for the difference in their susceptibility to diabetes.

SUMMARY

Fig 4 shows a summary of the cellular composition of BBDR/Ed and BBDP/Ed peripheral blood and cervical lymph node compared to PVG controls. The overall sizes of the boxes represent the yield of cells from each type of rat and the relative proportion of each cell type present is indicated by the way the box is subdivided. Fig 4a (the composition of blood) indicates clearly that although both the lines held at Edinburgh have lymphopenia and an overall reduction in peripheral blood leucocytes, they have numerically more monocytes and NK cells than non-lymphopenic control animals. It is possible that cells within these populations could be pathogenic and further analysis needs to be done in order to assess this. It is significant to note that in spite of their lymphopenia (which results in every lymphocyte cell population being underrepresented) the BB/Ed rats have significantly elevated numbers of these two non-lymphocyte cell types. Fig 4b shows the cellular composition of the lymph nodes from the rats analysed. It is clear that the lymphopenia results in a great reduction in the yield of leucocytes and because the T cell compartment is so severely affected (with no CD8+ Tcells at all) there is an increase in the proportion of B cells present.

CONCLUSION

The haplotyping experiment confirmed that both the BBDR and the BBDP rats from Edinburgh were carrying the permissive haplotype for diabetes in this model (namely the RT1ᵘ haplotype).The analysis of the cellular composition of periheral lymphoid compartments revealed that the BBDP from Edinburgh was displaying the severe lymphopenia and so was like all other high incidence BBDP colonies so far described. Surprisingly the BBDR rats from Edinburgh also displayed the lymphopenia and the two lines were indistinguishable from each other with respect to the analyses shown here. In all other colonies of BB rats the BBDR rats are non-lymphopenic and it has been thought that this trait, in association with the RT1ᵘ haplotype at class II always results in IDDM on this background.The BBDR/Ed subline carries two of the known susceptibilty genes for IDDM in this model, (namely the RT1ᵘ

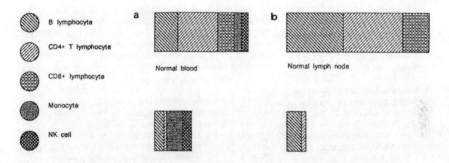

Fig 4. (a) and (b) Illustrate the cellular composition of the stated tissues.

Table 1. Reagents Used in This Study.

Name	Specificity	Ref
MRCOX8	CD8 (CD8+ T cells/NK cells)	26
MRCOX19	CD5 (T cells)	27
MRCOX38	CD4 (CD4+ T cells/monocytes)	28
MRCOX43	Some monocytes/ Some vascular endothelium	32
R73	TCR2 (T cells)	29
RG7/9.1	IgG k light chain (some B cells)	30
MARM4	Rat IgM (some B cells)	31
W3/13	CD43 (T cells / monocytes/neutrophils)	33
3.2.3	NK1 (NK cells)	34

159

haplotype and the gene for lymphopenia) it seems that in the BBDR rats from Edinburgh there may ,
be another susceptibility gene involved since they do not suffer from IDDM as would be predicted. We
are in pursuit of the genetic difference between the two lines of BB rats from Edinburgh.

ACKNOWLEDGEMENTS

I would like to thank Nigel Miller for his expert help with the flow cytometry. This work is supported
by the British Diabetic Association

References

1.   Chappel C.I and Chappel W.R, 1983. The discovery and development of the BB rat colony: an
        animal model of spontaneous diabetes mellitus. *Metabolism* 32 (suppl 1) p 8-10.
2.   Nakhooda A F, Like A A, Chappel C I, Murray F T, and Marliss E B, 1977. The spontaneously
        diabetic Wistar rat: metabolic and morphologic studies. *Diabetes* 26:100-112.
3.   Walker R, Bone A J, Cooke A and Baird J D, 1988. Distinct macrophage subpopulations in the
        pancreas of prediabetic BB/E rats: possible role for macrophages in the pathogenesis of IDDM.
        Diabetes 37 :1301-1304.
4.   Cooke A,1990. An overview on possible mechanisms of destruction of the insulin-producing
        beta cell. *Current topics in microbiology and immunology* 164 :125-137.
5.   Metroz Dayer M D, Mouland M, Brideau C, Duhamel D and Poussier P, 1990. Adoptive transfer
        of diabetes in BB rats induced by CD4+ lymphocytes. *Diabetes* 39:928-932.
6.   Like A A, Kislauskis E, Williams R M and Rossini A A, 1982. Neonatal thymectomy prevents
        spontaneous diabetes mellitus in the BB/W rat. *Science* 216:644-646.
7.   Bone A J, Walker R, Varey A M, Cooke A and Baird J D, 1990. Effect of cyclosporin on
        pancreatic events and development of diabetes in BB/Edinburgh rats. *Diabetes* 39:508-513.
8.   Jackson R, Rassi N, Crump T, Haynes B and Eisenbarth G S, 1981,. The BB diabetic rat.
        Profound T cell lymphocytopenia. *Diabetes* 30:887-889.
9.   Bellgrau D, Nagi A, Silvers W K, Markmann J F and Barker C F, 1982, Spontaneous diabetes in
        BB rats: evidence for a T cell dependent immune response defect. *Diabetologia* 23:359-364.
10.   Elder M and Maclaren N K,1983, Identification of profound T lymphocyte immunodeficiencies
        in the spontaneously diabetic BB rat. *J.Immunol.* 130:1723-1730.
        Woda B A, Like A, Padden C and McFadden M, 1986, Deficiency of phenotypic cytotoxic
        suppressor T lymphocytes in the spontaneously diabetic BB rat. *J I* vol 136 p 856-859.
12.   Greiner D, Handler E S, Nakano K, Mordes J P and Rossini A A, 1986, Absence of the RT6 T
        cell subset in diabetes prone BB/W rats. *J.Immunol.* 136:148-151.
13.   Thiele H G, Koch F Hammanand and Arndt R 1986. Biochemical characterisation of
        the T cell alloantigen RT6.2. *Immunology* 59:195-202
14.   Greiner D L, Mordes J P, Handler E S, Angellino M, Nakamura N and Rossini A A, 1987.
        Depletion of RT6.1+ T lymphocytes in diabetes resistant BioBreeding (BB/W) rats. *J. Exp.Med*
        166:461-475.
15.   Cormier J N, Handler E S, Mordes J P and Rossini A A, 1989. silica fails to prevent diabetes in
        RT6 depleted resistant (DR) BB rats or adoptive transfer of diabetes in diabetes prone (DP)
        BB rats. *Clin. Res.* 37 no.3 (meeting abstract)
16.   Colle E, Guttman R.D, Seemayer T.A 1981, Spontaneous diabetes mellitus syndrome in the rat.
        1. Association with the major histocompatibilty complex of the rat. *J.Exp.Med.* 154
        :1237-1242.
17.   Jackson R, Buse J B, Rifai R, Pelletier D, Milford e L, Carpenter C B, Eisenbarth G S and
        Williams M, 1984, Two genes required for diabetes in BB rats. Evidence from cyclical
        intercrosses and backcrosses. *J.Exp.Med* 159:1629-1636.
18.   Kloeting I and Stark O, 1987, Genetic studies of IDDM in BB rats: the incidence of diabetes in
        F2 and first backcross hybrids allows rejection of the recessive hypothesis. *Exp. Clin.
        Endocrinol.*'89 :312-318.
19.   Gunther E, Kiesel M, Kolb H, Krawczak M, Rothermel E and Wurst W, 1991, Genetic analysis
        of susceptibility to diabetes mellitus in F2 hybrids between DPBB and various MHC
        recombinant congenic rat strains. *J. Autoimmunity* 4 :543-551.
20.   Markholst H, Eastman S, Wilson D, Adreason B. E, and Lernmark A, 1991, Diabetes
        segregates as a single locus in crosses between inbred BB rats, prone or resistant to diabetes.
        J.Exp.Med. 174 :297-300.
21.   Colle E, Guttmann R D and Fuks A, 1986, Insulin dependent diabetes mellitus is associated with
        genes that map to the right of the class I RTI.A locus of the major histocompatibility complex of
        the rat. *Diabetes* 35 :454-458
22.   Colle E, Guttman R D, Fuks A, Seemayer T A and Prud'homme G J, 1986, Genetics of the
        spontaneous genetic syndrome. Interaction of MHC and non-MHC associated factors. *Mol. Biol.
        Med* 3:13-23.

23. Guttmann R D, Colle E, Michel F and Seemayer T, 1983, Spontaneous diabetes mellitus syndrome in the rat. II. T lymphopenia and its association with clinical disease and pancreatic lymphocytic infiltration. *J.Immunol.* 130:1732-1735.

24. Butcher G W, 1987, A list of monoclonal antibodies specific for alloantigens of the rat. *J. Immunogenetics* 14 :163-176.

25. Davidson W F and Parish C R, 1975, A procedure for removing red cells and dead cells from lymphoid cell suspensions. *J.Immunol. Methods* 7 :291-300.

26. Brideau R J, Carter P B, McMaster W R, Mason D W and Williams A F, 1980, Two subsets of rat lymphocytes defined with monoclonal antibodies. *Eur. J. Immunol.* 10 :609-615.

27. Dallman M J, Mason D W and Webb M, 1982. The roles of host and donor cells in the rejection of skin allografts by T cell depleted rats injected with syngeneic T cells. *E.J.I* 12;511-518

28. Jeffries W A, Green J R and Williams A F, 1985, Authentic T helper CD4 (W3/25) antigen on rat peritoneal macrophages. *J. Exp. Med* 162:117-127.

29. Hunig T, Wallney H J, Hartley J K, Lawetzky A and Tiefenthaler G, 1989, A monoclonal antibody to a constant determinant of the rat T cell antigen receptor that induces T cell activation. Differential reactivity with subsets of immature and mature T lymphocytes. *J. Exp. Med* 169:73-86.

30. Springer T A, Bhattacharya A, Cardoza J T and Sanchez-Madrid F, 1982, Monoclonal antibodies specific for rat IgG1 and IgG2a and IgG2b subclasses and kappa chain monotypic and allotypic determinants: reagents for use with rat monoclonal antibodies. *Hybridoma* 1:257-273.

31. Malache J M, Manouvriez P and Bazin H, 1990, List of monoclonal antibodies. Chapter in: Rat hybridomas and monoclonal antibodies. (Ed) H.Bazin. CRC Press, Boca Raton, Florida.

32. Robinson A P, White T M amd Mason D W, 1986. MRCOX43: a monoclonal antibody which reacts with all vascular endothelium in the rat except that of brain capillaries. *Immunology* 57:231-237.

33. Williams A, Galfre G and Milstein C, 1977. Analysis of cell surfaces by xenegeneic myeloma hybrid antibodies: Differentiation of rat lymphocytes. *Cell* 12 :663-673

34. Chambers W H, Vujanovic N L, Deleo A B, Olszowy M W, Herberman R B and Hiserodt J C, 1989. Monoclonal antibody to a triggering structure expressed on rat natural killer cells and adherent lymphokine activated killer cells. *J.Exp. Med* . 169:1373-1389

35. Like A A, Biron C A, Weringer E S, Brayman K, Sroczynski E and Guberski D L, 1986. Prevention of diabetesin BioBreeding Worcester rats with monoclonal antibodies that recognize T lymphocytes or natural killer cells. *J.Exp. Med* 164:1145-1159.

36. Rothe H, Fehsel K and Kolb H, 1989. Aberrant production of TNFα by macrophages from diabetes prone BB rats. Abstract (p 26) in *Lessons from animal diabetes, 3rd international workshop* in Israel.

KEYWORDS:
BB RAT/ IDDM/ TOLERANCE/ LYMPHOPENIA/ MONOCYTE/ NK CELL/ SUSCEPTIBILTY GENE.

ABSRACT:
We describe an analysis of the genetic characteristics of the BB rat colony from Edinburgh. This colony consists two lines: one with a high incidence of IDDM (BB-DP) and one with no disease (BB-DR). We have found that in accordance with published data, the BB-DP line displays a severe lymphopenia with no CD8+ T cells and depressed numbers of all other circulating lymphocytes. Interestingly, the BB-DR subline from Edinburgh also displays this trait. We have identified the cell populations present in the two lines and can find no phenotypic difference between them. One feature of the lymphopenia does seem to be enlarged populations of NK cells and monocytes, although the significance of this is unclear. This is the first time that lymphopenia has been seen in a diabetes resistant subline of BB rats.Both the lines held in Edinburgh carry the RT1ᵘ haplotype at the class I and the class II loci and they therefore both carry two known susceptibility genes for IDDM in this model. The genetic difference between the two lines remains to be identified and is under investigation.

17

RT6, AN UNUSUALLY POLYMORPHIC T-CELL SPECIFIC MEMBRANE PROTEIN

LINKED TO TYPE I AUTOIMMUNE DIABETES IN RATS AND MICE

Friedrich Haag, Friedrich Koch-Nolte, Christiane
Hollmann, Heinz Günter Thiele

Dept. of Immunology
Universitätskrankenhaus Eppendorf
Hamburg, FRG

ABSTRACT

RT6 is a 25-30kd phosphatidylinositol-linked membrane protein of
T cells. Recently, we have obtained cDNA clones encoding rat and
mouse RT6 allelic variants. The RT6 antigens are encoded by a
single gene locus in rats and by two closely linked gene loci in
mice. RT6 allelic gene products of rats and mice are unusual in
that they are highly polymorphic. Conspicuous features of the
mutations suggest that they may be of functional significance.

In both rats and mice, RT6 gene expression evidently is
restricted to the postthymic T cell lineage. Data from the rat
system indicate that the number of RT6[+] cells increases slowly
during postnatal ontogeny.While recent thymic migrants are RT6[-],
most resting T cells in the adult rat are RT6[+]. Activation of T
cells and *in vitro* cultivation lead to loss of the RT6[+]
phenotype.

Interestingly, two indepedent animal models of autoimmune type 1
diabetes - the dpBB rat and the NOD mouse - show an inherent
defect in RT6 expression.

RT6 is a marker for mature, resting, peripheral T cells

The RT6 alloantigenic system in the rat comprises at least two
known alleles, RT6[a] and RT6[b]. It was originally discovered
because of its potent immunogenicity in the allogeneic context.[1-3]
Cross-immunizing different strains of rats with peripheral T
lymphocytes yielded antisera containing high titers of allotype-
specific antibodies that discriminate between the two known
allelic gene products, designated RT6.1 and RT6.2.

T Lymphocytes: Structure, Function, Choices, Edited by F. Celada
and B. Pernis, Plenum Press, New York, 1992

Immunoprecipitation experiments revealed protein products with apparent molecular weights of 25 -30 kd. The RT6 antigens are unusual membrane proteins in that they occur in non-glycosylated (RT6.2) or in non-glycosylated and differently glycosylated (RT6.1) variants.[4] All forms of RT6 are anchored to the cell membrane by covalent linkage to a phosphatidylinositol (PI) moiety.[5] Treatment of cells with RT6-specific antisera reportedly is mitogenic,[6] similar to the case of other PI-linked proteins.[7]

In the rat, the expression of RT6 on the cell surface is restricted to a late stage of postthymic T cell development, following a pattern reciprocal to that of Thy-1. The majority of mature peripheral T cells, of both the CD4+ as well as the CD8+ subsets, carry the phenotype RT6+/Thy-1-. Thymic lymphocytes and recent thymic migrants are RT6-/Thy-1+.[8] Recent studies have reported very high levels of RT6 expression by intraepithelial lymphocytes in the gut.[9] Available evidence indicates that RT6 is not expressed by other hematopoietic cells nor by cells of other tissues.[8,10] RT6 gene expression appears to be under strict regulatory control. In vitro cultivation as well as activation of T cells regularly leads to loss of the RT6+ phenotype.[11]

Northern blot and PCR analyses show a similar pattern of RT6 gene expression in mouse and rat.

The RT6 system is highly polymorphic

Our laboratory has recently reported the molecular cloning of cDNAs coding for DA rat RT6.2[12,13] and Lewis rat RT6.1.[14] Sequence analyses revealed that RT6.1 and RT6.2 are complex allotypes, differing in 10 out of the 226 amino acids which make up the native protein.[15] The majority of these substitutions are non-conservative with respect to the biochemical properties of the amino acids involved, i.e. 7 of the 10 result in a charge change (Table 1). The differences are not randomly distributed throughout the length of the molecule, but appear in two main clusters. One such cluster of three consecutively mutated amino acids encodes a single potential glycosylation site (asn-lys-ser) in RT6.1 not present in RT6.2.

Southern blot analyses revealed restriction fragment length variants (RFLVs) between rat strains that correspond to the serological typing of the strains as RT6a or RT6b. Restriction and sequence analyses of PCR amplification products from genomic DNA confirm that RT6a and RT6b are two alleles of a single gene locus.[12,13] The results of these and serological studies (G. Butcher, personal communication) suggest the possible occurrence of additional RT6 (sub-)allelic variants in other rat strains. Using a pair of synthetic oligonucleotides containing the AUG start codon and the TAG stop codon, respectively, we have amplified DNA encoding the murine homologue of RT6 (designated Rt-6) from Balb/c spleen cDNA by PCR methods.[16] As in the rat, Southern blot analysis reveal RFLVs between different mouse strains with many different restriction endonucleases. In contrast to the rat, however, the hybridization pattern on these blots indicate the presence of two homologous genes.[17] Sequence information obtained from PCR amplification products of genomic

Table 1

Amino acid substitutions between Lewis rat RT6.1 and DA rat RT6.2

residue No. from N-terminus	RT6.1	RT6.2	comment
34	lys	gln	charge change
39	lys	met	charge change
40	ser	asn	creates glycosylation site
41	glu	ala	charge change
61	met	arg	charge change
83	gly	ala	
153	thr	lys	charge change
158	pro	gln	
164	asp	his	charge change
184	tyr	arg	charge change

DNA and cDNA confirm the presence of two distinct Rt-6 genes with sequence identity of approximately 80%. Linkage analyses indicate that these loci are closely linked (see below). At least three highly polymorphic Rt-6 haplotypes can be distinguished in inbred strains of mice. Furthermore, as in the rat, the Southern blot data suggest the presence of additional suballelic variants. It remains to be determined whether the allelic polymorphism observed for murine Rt-6 by RFLV analysis also translates into extensive polymorphism on the amino acid level as has been observed for the rat.

Searches in data bases have not revealed any significant homologies of RT6 to known proteins. In particular, detailed analyses indicate that the RT6 antigens are not members of the immunoglobulin superfamily (A. Williams, personal communication).

The RT6 gene is located in a region of high synteny homology between rat, mouse and humans

The gene for rat RT6 has previously been mapped to linkage group I (chromosome 1) between the loci for hemoglobin β (Hbb) and albinism (c).[18] Probing of Southern blots from recombinant inbred strains of mice showed that the two Rt-6 genes are closely linked on chromosome 7 and also localized between the Hbb and c gene loci.[17] This region shows a high linkage synteny homology to human chromosome 11p.

Absence of the RT6 subset of peripheral lymphocytes is associated with autoimmune type I insulin-dependant diabetes mellitus (IDDM)

In 1986 it was observed that the diabetes prone Bio-Breeding (dpBB) rat, an animal model for human autoimmune type I IDDM, has an inherent defect in the generation of RT6+ cells.[19]

Subsequently, it was reported that *in vivo* depletion of RT6[+] T cells in young diabetes resistant BB (drBB) rats leads to development of diabetes in high incidence,[20] whereas disease in dpBB rats may be prevented by transfusion of RT6[+] cells. These findings may point to an important regulatory function exercised by a subset of RT6[+] T cells.[21] Interestingly, Northern Blot analysis of RNA from spleens of young, prediabetic NOD (non-obese diabetic) mice, another animal model for human IDDM, shows reduced transcription for Rt-6 in comparison to non-diabetes prone NON mice of equal age.[17]

Molecular cloning of the human RT6 homologue

Using degenerate primers corresponding to conserved regions of rat and mouse RT6 we have recently succeeded in amplifying a fragment of the human RT6 gene. This will enable us to study a possible linkage between type I autoimmune diabetes and a defect in RFT6 expression also in the human and to identify alleles that may correlate with disease susceptibility. Corresponding experiments are underway in our laboratory.

REFERENCES

1. G.W.Butcher and J.C. Howard, An alloantigenic system on rat peripheral T cells, Rat News Letter 1:12 (1977)
2. C.W.DeWitt and M.McCullough, AgF: serological and genetic identification of a new locus in the rat governing lymphocyte membrane antigens, Transplantation 19:310 (1975)
3. H.-G.Thiele, R.Arndt, R.Stark & K.Wonigeit, Detection and partial molecular characterization of the rat T-lymphocyte surface protein L_{21} by allo- (anti-RT Ly 2.2) and xeno-(anti-RT.LN-Ly [Ig-]) sera, Transpl. Proc. 11:1636 (1979)
4. F.Koch, A.Kashan & H.-G.Thiele: The rat T cell marker RT6.1 is more polymorphic than its alloantigenic counterpart RT6.2, Immunol. 65:259 (1988)
5. F.Koch, H.-G.Thiele & M.Low, Phosphatidylinositol is the membrane anchoring domain of the rat T cell antigens RT6.2 and Thy-1, Transpl. Proc. 19:3140 (1987)
6. K.Wonigeit and R.Schwinzer, Polyclonal activation of rat T lymphocytes by RT6 alloantisera, Transpl. Proc. 19:29 (1987)
7. M.G.Low and A.R.Saltiel: "Structural and functional roles of glycosyl-phosphatidyl-inositols in membranes." Science 239:268 (1988)
8. H.-G.Thiele, F.Koch & A.Kashan, Postnatal distribution profiles of Thy-1 and RT6.2[+] cells in peripheral lymph nodes of DA rats, Transpl. Proc. 19:3157 (1987)
9. J.Fangmann, R.Schwinzer, M.Winkler, and K.Wonigeit, Unusual phenotype of intestinal epithelial lymphocytes in the rat: Predominance of T-cell receptor α/β^+/CD2[-] cells and high expression of the RT6 alloantigen, Eur. J. Immunol. 21:753 (1991)
10. H.-G.Thiele, F.Koch, F.Haag, and W.Wurst, Evidence for normal thymic export of lymphocytes and an intact RT6[a]

gene in RT6-deficient diabetes prone BB rats, Thymus
13:137 (1989)

11. F.Koch, A.Kashan and H.-G.Thiele, Production of a rat T cell
    hybridoma that stably expresses the rat T cell
    differentiation marker RT6.2, Hybridoma 7:341 (1988)

12. F.Koch, F.Haag, A.Kashan and H.-G.Thiele, Construction of a
    rat hybridoma cDNA expression library and isolation of
    a cDNA clone for the T cell differentiation marker
    RT6.2, Immunol. 67:344 (1989)

13. F.Koch, F.Haag, A.Kashan and H.-G.Thiele, Primary structure
    of rat RT6.2: a non-glycosylated PI-linked surface
    marker of postthymic T cells, Proc. Natl. Acad. Sci
    USA 87:964 (1990)

14. F.Haag, F.Koch and H.-G.Thiele, Nucleotide and deduced amino
    acid sequence of the rat T cell alloantigen RT6.1,
    Nucl. Acids Res. 18:1047 (1990)

15. F.Haag, F.Koch, and H.-G.Thiele, Polymorphism between rat T-
    cell alloantigens RT6.1 and RT6.2 is based on multiple
    amino acid substitutions, Transpl. Proc. 22:2541
    (1990)

16. F.Koch, F.Haag, and H.-G.Thiele, Nucleotide and deduced
    amino acid sequence for the mouse homologue of the rat
    T-cell differentiation marker RT6, Nucl. Acids Res.
    18:3636 (1990)

17. M.Prochazka, H.R.Gaskins, E.H.Leiter, F.Koch-Nolte, F.Haag,
    and H.-G.Thiele, Chromosomal localization, DNA
    polymorphism,and expression of Rt-6, the mouse
    homologue of rat T-lymphocyte differentiation marker
    RT6, Immunogenetics 33:152 (1991)

18. G.W.Butcher, S. Clarke, and E.M.Tucker, Close linkage of
    peripheral T-lymphocyte antigen A (PtA) to the
    hemoglobin variant Hbb on linkage group I of the rat,
    Transpl. Proc. 11:1629 (1979)

19. D.L.Greiner, E.S.Handler, K.Nakano, J.P.Mordes &
    A.A.Rossini, Absence of the RT6 T-cell subset in
    diabetes-prone BB/W rats, J. Immunol. 136:148 (1986)

20. D.L.Greiner, J.P.Mordes, E.S.Handler, M.Angelillo,
    N.Nakamura & A.A.Rossini, Depletion of RT6.1[+] T-
    lymphocytes induces diabetes in resistant
    Biobreeding/Worcester (BB/W) rats, J. Exp. Med.
    166:461 (1987)

21. D.L.Greiner, J.P.Mordes, M.Angelillo, E.S.Handler,
    C.F.Mojcik, N.Nakamura, and A.A.Rossini, Role of
    regulatory RT6[+] T-cells in the pathogenesis of
    diabetes mellitus in BB/Wor rats, in: "Frontiers in
    diabetes research: Lessons from animal diabetes II,"
    E.Shafrir and A.E.Renold, ed., John Libbey, London
    (1988)

18

ALTERNATIVE PATHWAYS OF SIGNAL TRANSDUCTION AFTER LIGATION OF THE TCR BY

BACTERIAL SUPERANTIGEN

Hsi Liu and Harvey Cantor

Department of Pathology, Harvard Medical School
Laboratory of Immunopathology, Dana-Farber Cancer Institute
Boston, MA

INTRODUCTION

The development of mature T-cell subsets can be divided into several critical steps beginning in the thymus and continuing in peripheral lymphoid tissues. Ligation of the surface $\alpha\beta$ receptor on $CD4^+8^+$ cells can result in survival (positive selection) (Benoist & Mathis, 1989) and further maturation along either the CD4/helper pathway or CD8/cytotoxic pathway. Clones that escape intrathymic deletion following TCR ligation by MHC/self peptide complexes on dendritic cells may migrate to the periphery, where subsequent TCR ligation can result in clonal anergy or in expression of specialized T-cell function (Schwartz, R. 1990; Whiteley et al, 1990). In the case of $CD4^+$ T_H cells, ligation of the TCR may lead to alternative functional phenotypes associated with different patterns of cytokine gene expression. The TCR-linked signalling events responsible for acquisition of these successive alternative phenotypes during the development of T-cells are not understood.

One explanation holds that occupation of the T-cell receptor (TCR) results in a single event which can yield multiple biological outcomes, depending on the composition of intracellular proteins present in T-cells at particular stages of differentiation. A second explanation is that expression of alternative phenotypes during T-cell development may reflect coupling of the TCR to functionally distinct signal transduction pathways. According to this model, the alternative phenotypes associated with positive and negative intrathymic selection, subset commitment and peripheral responsiveness may reflect different forms of TCR ligation coupled to distinct pathways of intracellular signal transduction.

Over the past several years, we have obtained evidence that two functionally distinct signal transduction pathways may be coupled to the TCR of the $CD4^+$ T_H subset (Patarca et al, 1991; Liu et al.,1991a,b). The two pathways are characterized by distinctive intracellular events and expression of different functional phenotypes. One signal transduction pathway that is coupled to TCR ligation by appropriate peptide/MHC complexes is marked by increased $[Ca^{2+}]_i$, PKC dependence, proliferation and conventional cytokine expression. A second pathway is coupled to ligation of certain TCR by bacterial toxins or Mls-1^a/MIV superantigens. This pathway is marked by a proliferative response that is not accompanied by increased $[Ca^{2+}]_i$, PKC dependence or release of IL-2 and is followed by clonal unresponsiveness. The following summarizes studies that have defined these two TCR-linked activation pathways.

Definition of an Alternative Signal Transduction Pathway in CD4⁺ T-cells
Following TCR Ligation by Retroviral or Bacterial Superantigen

The possibility that the interaction of the TCR with peptide antigen and
with superantigen might have distinct functional consequences came from studies
of a $V_\beta 6^+$ triple-reactive T-cell clone that responds to Mls-1ᵃ, OVA peptide and
alloantigen (H-2ᵇ) (Friedman et al., 1987; Patarca et al. 1991). Ligation of
the TCR of this clone by H-2 alloantigen or peptide antigen resulted in a
proliferative response accompanied by vigorous gamma-IFN gene expression. In
contrast, ligation of the TCR by Mls-1ᵃ (MTV-7) resulted in equally strong
proliferative responses that were not accompanied by detectable changes in
gamma-IFN gene expression (Patarca et al., 1991) nor by a detectable increase in
$[Ca^{2+}]_i$ (Liu et al, 1991a). In the experiments summarized below, we attempted
to directly characterize these putative alternative signalling pathways using a
more well-defined class of superantigen, the Staphylococcal enterotoxins (SE).

To determine whether the proliferative response to the two types of ligand
was accompanied by distinct intracellular signals, we measured $[Ca^{2+}]_i$ which,
along with tyrosine phosphorylation, represents the earliest intracellular
change associated with cellular activation (Ashwell, 1990). Stimulation of
murine CD4⁺ T-cell clones with conventional protein antigen or anti CD3
antibody resulted in an initial sharp rise of $[Ca^{2+}]_i$ followed by a persistent
plateau. In contrast, stimulation of the two clones with bacterial
superantigen (SEB) or retroviral superantigen (Mls-1/Mtv-7) using
concentrations sufficient to trigger strong proliferative responses, was not
accompanied by a detectable rise in $[Ca^{2+}]_i$ (Liu et al., 1991a).

We also asked whether the two types of T-cell ligand provoked functionally
distinct signalling pathways according to cytokine production. Stimulation of
T-cell clones O3 and Ar5 by peptide ligand resulted in strong IL-2 and IL-3
responses that were in the range of 50-200 U/culture. By contrast, the
cytokine response to SEB was barely measureable (<0-5 U/culture) despite the
fact that the proliferative response to the two ligands was similar (Liu et al,
1991a,b). We also examined the effects of mixtures of the two ligands upon
cytokine production. We found that mixtures of the two types of ligand
displayed a strongly synergistic interaction in T-cell activation, as judged by
production of both IL-2 and IL-3. This finding is reminiscent of earlier
observations that the interaction of T-cells with Mls-1⁺ accessory cells
enhance the response to conventional (peptide) antigen and may thus act as a
coligand in this response (Janeway et al., 1983). The observed synergism
reflected a cooperative effect of independent signalling pathways rather than
increased crosslinking of the TCR by both ligands at the cell surface because
synergy was evident when each ligand was presented on different antigen
presenting cells (APC) (Liu et al, 1991a). Although additional studies are
needed to further define and distinguish the series of intracellular
biochemical events initiated after TCR ligation by these two classes of ligand,
these data suggest that the two activation pathways are likely to be
functionally distinct rather than alternatives of the same pathway.

TCR V_β Usage by CD4⁺ T-cell Clones that Express Distinct Signalling Pathways
after SEB Ligation

A key element in the definition of superantigen recognition involves
selective usage of V_β TCR elements. Therefore, studies using a panel of T-cell
clones that express defined V_β elements including $V_\beta 2$, 3, 6 and 8 were
performed to define V_β expression on murine T_H clones which respond to
bacterial toxins using the conventional or alternative activation pathways.

$V_\beta 2^+$ and $V_\beta 6^+$ clones, which have not been previously associated with
reactivity to SEB (Callahan et al, 1990; Yagi et al, 1990; White et al, 1989),
displayed strong proliferative responses to SEB. It is unlikely that the
response of these clones reflects recognition of a contaminant in the SEB
preparation (Callahan et al, 1990; Herman et al, 1991), because neither $V_\beta 2$-
nor $V_\beta 6$-containing TCR have been associated with the response of T-cells to any

of the other Staphylococcal toxins which might contaminate the SEB preparation (Herman et al., 1991).

The proliferative response of these clones to SEB did not reflect a 'back-reaction' in which the $V_\beta 6^+$ or $V_\beta 2^+$ T-cell clones respond to cytokines produced by irradiated splenic cells in the feeder population since incubation of the $V_\beta 6^+$ clone O3 with SEB using spleen cells from BALB/c nu/nu donors as APC resulted in vigorous SEB-dependent proliferative responses by the CD4$^+$ $V_\beta 6^+$ T-cell clone (Liu et al, 1991b).

Previous studies showing highly restricted toxin-reactivity of T-cell hybridomas and T-cell clones usually measured cytokine release rather than proliferation (Yagi et al., 1990; Callahan et al., 1990). Indeed, when this assay was used to monitor cell activation, only the response of the $V_\beta 3^+$ clone to SEB was accompanied by substantial IL-2 production (Table I and Liu et al, 1991b). In contrast, the strong proliferative responses of the $V_\beta 6^+$ and $V_\beta 2^+$ clones to SEB were not accompanied by significant release of IL-2. Nonetheless, in view of the unexpected nature of this response and V_β usage of

Table I. IL-2 Production after Toxin Stimulation of T-cell Clones Expressing Different V_β TCR Elements.

Clone	V_β Expression	Ligand in Culture	IL-2 (U/ml)
BC.4	3	none	-
		SEB	+++
		Cow insulin	++
O3	6	none	-
		SEB	-
		OVA	+++
		Mls-1[a]	-
Ar5	2	none	-
		SEB	-
		Ar-OVA	+++

The indicated clones were stimulated with SEB (10 µg/ml) or their specific protein ligands (cow insulin, 20 µg/ml; OVA, 10 µg/ml or Ar-OVA, 10 µg/ml). Supernatant fluids were tested for IL-2 activity as described (Liu et al., 1991a) and those values are represented according to the following scale: 0-1= -; 2-20= +; 21-40= ++; 41-300= +++.

these CD4$^+$ T-cell clones to SEB, we performed extensive studies to document this response reflected ligation of the TCR by SEB associated with class II products on APC.

ALTERNATIVE PATHWAYS OF SIGNAL TRANSDUCTION

A. Requirement for Class II Products on APC

The clones responded to SEB associated with APC of BALB/c and other strains expressing I-E, including B10.A2R (I-A^k, I-E^k) but not B10.A4R (I-A^k,

$I-E^b$) (Table II). Moreover, anti-$I-A^d$ inhibited the OVA response of O3 cells, but had no effect on the SEB response, while antibody that also recognized $I-E^d$ (M5.114) was required to inhibit the response to the latter ligand (Table II).

B. Interaction of SEB with TCR-containing Alternative V_β elements

The proliferative response of the $V_\beta6^+$ O3 clone to SEB was almost completely inhibited by $F(ab')_2$ fragment of anti $V_\beta6$ antibody while this antibody had no effect on the proliferative response of this clone to IL-2 (Liu et al., 1991b). It remained formally possible that SEB indirectly stimulates T-cells by stimulating macrophages and B-cells via I-E receptors to express new surface ligands and/or macrophage cytokines that cause T-cell proliferation. Grossman et al (1990) have shown that reduction and alkylation of SEB destroys the T-cell epitope of this molecule, but does not affect its ability to bind to class II products and activate macrophages. Reduced/alkylated SEB (SEB_{RA}) failed to stimulate $V_\beta2^+$ and $V_\beta6^+$ T-cell clones to incorporate ^{3}H-thymidine (Liu et al., 1991b).

Thus, the proliferative response of these murine T-cell clones reflects ligation of the TCR by SEB associated with class II products since (a) the $F(ab')_2$ fragment of anti $V_\beta6$ blocks the proliferative response of the $V_\beta6$ clone O3 to SEB but not IL-2, (b) destruction of the T-cell epitope but not the class II binding epitope prevents the proliferative response, and (c) the response depends on expression of I-E (but not I-A) products displayed on either B-cell lines or T-depleted spleen cells as a source of APC.

Table II. Requirement for Class II Expression on APC in the Response of O3 and Ar5 to SEB and Conventional Antigen.

Clone	Source of APC	I-A	I-E	^{3}H-T Incorporation (Ar)-OVA	SEB	none
O3	BALB/c	d	d	+++	+++	-
	B10.A2R	k	k	-	++	-
	B10.A4R	k	-	-	-	-
Ar5	BALB/c	d	d	+++	+++	-
	B10.A2R	k	k	-	+++	-
	B10.A4R	k	-	-	-	-

The indicated clones were stimulated by the bacterial toxin SEB or by conventional antigen (OVA for O3; Ar-OVA for Ar5) using $5x10^5$ irradiated (2000R) spleen cells from different inbred mouse strains. Thymidine incorporation 24-36 hours after initiation of each response is shown and values are represented according to the following scale (cpm x 10^{-3}) : 0-10= -; 11-50= +; 51-100= ++; 101-400= +++.

C. Development of Clonal Unresponsiveness

The ability of Staphylococcal toxins to stimulate large numbers of T-cells has led to their designation as superantigens and it is likely that excessive T-cell activation by superantigens is responsibel for the toxic effects of these bacterial products (Marrack & Kappler, 1990; Janeway et al., 1989). However, expression of superantigens by diverse species of bacteria including Mycoplasma (Cole et al., 1989), Pseudomonas (Misfeldt et al., 1990) and Streptococcus (Kotb et al., 1989), suggests the possibility that these proteins may represent a successful microbial strategy for subversion of the host immune response.

Although in vivo administration of Staphylococcal enterotoxin B (SEB) may be followed by SEB-specific anergy (Rellahan et al., 1990; Kawabe & Ochi, 1990), this type of restricted unresponsiveness is not likely to affect the host antibacterial response. A more relevant observation comes from early studies indicating that in vivo administration of SEB may be followed by inhibition of both humoral and cellular immune responses (Pinto et al., 1978). However, the cellular basis of generalized immunological inhibition by SEB is not understood.

In the above studies, we found that SEB may interact with the TCR of a larger fraction of T-cells than has been previously suggested, resulting in an alternative activation pathway marked by proliferation but not cytokine production (Patarca et al., 1991; Liu et al., 1991a). We asked whether this activation pathway also might affect subsequent responsiveness of T-cell clones to TCR ligation.

O3 cells were stimulated by SEB or by their respective conventional peptide ligands and BALB/c adherent spleen cells as APC. Twenty-four hours later, the cells were retrieved, transferred to cultures containing fresh APC and different concentrations of OVA. Measurement of IL-2 or IL-3 production showed that cells that had initially been stimulated with SEB failed to produce significant levels of cytokine upon subsequent stimulation by peptide ligand and fresh APC (Liu et al., 1991b). By contrast, cells that had been initially stimulated either with conventional peptide ligand/APC or with APC alone produced high levels of these cytokines. SEB-stimulated O3 cells also exhibited reduced responsiveness when restimulated with anti-CD3 or anti-$V_\beta6$, as judged by incorporation of ^{3}H-thymidine (Liu et al, 1991b). In the above experiments, adherent cells from the spleens of syngeneic euthymic (BALB/c) mice were used as a source of APC for the initial stimulation. We tested the possibility that although T-cells in the splenic adherent cell population used as APC did not produce detectable levels of IL-2 in the supernatant fluids and little or no IL-2 according to co-cultivation assays, they might secrete cytokines that contributed to the development of T-cell unresponsiveness following stimulation by SEB. Incubation of O3 cells with SEB and adherent cells from (BALB/c) nu/nu mice resulted in substantially reduced cytokine responses to subsequent stimulation with OVA (Liu et al., 1991b). Since these APC did not stimulate HT-2 cells in the presence of SEB, it is unlikely that cytokines produced by T-cells in the APC population contributed significantly to the expression of the unresponsive phenotype.

The findings described above suggested that in vivo administration of SEB might be followed by polyclonal T-cell unresponsiveness rather than anergy limited to $V_\beta8^+$ T-cells as reported previously (Rellahan et al., 1990). To test this hypothesis, we injected mice with 50 ug/ml of SEB or with PBS. Seven days later, T-cells from the former mice exhibited reduced responses to antibodies to $V_\beta2$, $V_\beta6$, $V_\beta8$, and $V_\beta17a$, as well as to anti CD3. In addition, a substantial reduction in the T-cell responses to both SEB and the structurally unrelated toxin TSST-1 was also apparent (Liu et al., 199b).

DISCUSSION

T-cells that express a restricted set of V_β elements have been reported to respond to SEB according to proliferation and cytokine production (Yagi et al., 1990; Callahan et al., 1990). The studies summarized here have begun to define a second category of SEB reactivity that includes T-cell clones that bear alternative V_β elements. The response of these clones to superantigens is characterized by proliferation, little or no detectable secretion of IL-2 or IL-3 and the development of clonal unresponsivess. This form of T-cell activation reflects ligation of the TCR by SEB because the response is inhibited by a F(ab')$_2$ anti V_β fragment and depends on presentation of SEB in association with I-E products on APC.

Previous evidence that reactivity to SEB is limited to T-cells bearing particular V_β elements is based mainly on IL-2 production (Yagi et al., 1990;

Callahan et al, 1990). Although we have confirmed that IL-2 production by clones bearing particular V_β elements (e.g., $V_\beta 3$, $V_\beta 8$) marks this reponse to SEB, we also find that the interaction of SEB with TCR containing alternative V_β elements (e.g., $V_\beta 2$, $V_\beta 6$) results in stimulation of ^{3}H-T incorporation without significant levels of cytokine expression.

The molecular basis of TCR recognition of SEB by clones bearing alternative V_β elements is not yet understood, although we have shown a requirement for an intact disulfide loop. Recent studies have indicated that (a) mutations of this disulfide loop reduce T-cell stimulatory activity without affecting class II binding or the overall configuration of the toxin (Grossman et al, 1990, 1991) and (b) exchanges of the loop regions between SEA and SEB do not affect the V_β usage of responder T-cells. These findings are consistent with the idea that the loop region may interact with a conserved portion of the V_β chain, while an independent region of the toxin may interact with sites expressed on particular V_β chains (e.g., $V_\beta 8$) (Choi et al., 1990). Additional studies using recombinant SEB proteins having mutations in the loop region and elsewhere are necessary to more precisely define the potential contribution of toxin subregions to the response described here.

The functional consequences of the interaction between SEB and TCR containing alternative V_β elements include an early proliferative response without substantial cytokine (IL-2/IL-3) production and the development of clonal unresponsiveness to subsequent antigenic stimulation by peptide ligands. Although ligation of T-cell clones by peptide ligand results in a rapid and sustained increase in $[Ca^{2+}]_i$, ligation of these T-cell clones by SEB is not accompanied by a detectable increase in $[Ca^{2+}]_i$ levels (Liu et al, 1991a). It is therefore unlikely that development of the unresponsive state following this form of activation is mechanistically similar to the development of clonal anergy to T-cell ligands presented in the absence of costimulatory signals (Schwartz R, 1990; O'Hehir & Lamb, 1990).

The extent of SEB-dependent sterile T-cell activation among populations of TCR-bearing clones is not known. Nonetheless, the findings summarized here indicate that the toxin interacts with a broader range of T-cell receptors than has been generally assumed and can cause a generalized hyporesponsiveness of T-cells after in vivo adminstration of SEB. This may be relevant to early studies suggesting that SEB can inhibit homograft reactivity and antibody responses to the T-cell-dependent antigen SRBC but not to the relatively T-independent antigen LPS (Pinto et al., 1978).

These observations also suggest a discordance between proliferation and the $T_H 1$ cytokine response following ligation of the TCR. One explanation for this dissociation is that TCR ligation may be coupled to distinct signalling pathways characterized by different intracellular biochemical events and patterns of gene expression (Liu et al., 1991a; Sussman et al., 1988; Mercep et al., 1989; O'Rourke et al., 1990). Although we have previously noted that the two potentially distinct activation pathways may differ with respect to changes in $[Ca^{2+}]_i$ immediately after stimulation by the two ligands (Liu et al., 1991a), examination of additional intracellular events such as PI hydrolysis and coupling to G proteins is necessary to further distinguish these two putative pathways.

The potential role of these activation pathways in expression of alternative phenotypes during intrathymic differentiation is also not addressed by these studies. Nonetheless, the apparent dissociation between proliferation and cytokine production also suggests that accurate evaluation of T-cell immune status should include measurements of both response parameters. Indeed, previous studies of heterogenous populations of CD4⁺ T-cells have indicated that T-cells which display efficient proliferative responses to antigen may not express detectable T-helper/effector activity (Tite et al., 1987; Peterson et al., 1983; Krzych et al., 1982). These experiments used whole populations of T-cells whereas we have separated the two relevant components of T-cell activation (proliferation and cytokine release) using the TCR of cloned populations of T_H cells.

Finally, these results may provide insight into the evolutionary basis for expression of superantigens by diverse bacterial species. Excessive cytokine production and associated in vivo toxicity may reflect previously described cytokine responses by a relatively restricted set of clones that express particular V_β elements (Marrack et al., 1990). However, T-cell unresponsiveness resulting from the interaction between superantigens and a conserved region of the TCR may be more relevant to the widespread expression of these interesting bacterial products.

REFERENCES

Ashwell, J. D., 1990, Genetic and mutational analysis of the T-cell antigen receptor, Ann. Rev. Immunol., 8:139.

Benoist, C., and Mathis, D., 1989, Positive selection of the T-cell repertoire: where and when does it occur? Cell, 58:1027.

Callahan, J. E., Herman, A., Kappler, J. W., and Marrack, P., 1990, Stimulation of B10.BR T cells with superantigenic staphylococcal toxins, J. Immunol., 144:2473.

Choi, Y., Herman, A., DiGiusto, D., Wade, T., Marrack, P. and Kappler, J., 1990, Residues of the variable region of the T-cell receptor ß-chain that interact with S. aureus toxin superantigens, Nature, 346:471.

Cole, B. C., Kartchner, D. and Weller, D., 1989, Stimulation of mouse lymphocytes by a mitogen for Mycoplasma arthriditis, J. Immunol., 142:4131.

Friedman, S., Sillcocks, D., and Cantor, H., 1987, Alloreactivity of an OVA-specific T-cell clone. I. Stimulation by class II MHC and novel non-MHC B-cell determinants, Immunogenetics, 26:193.

Grossman, D., Cook, R. G., Sparrow, J. T., Mollick, J. A., and Rich, R. R., 1990, Dissociation of the stimulaotry activities of staphylococcal enterotoxins for T cells and monocytes, J. Exp. Med., 172:1831.

Grossman, D., Van, M., Mollick, J. A., Highlander, S. K. and Rich, R. R., 1991, Mutation of the disulfide loop in Staphylococcal enterotoxin A: Consequences for T cell recognition, J. Immunol., 147:3274.

Herman, A., Kappler, J. W., Marrack, P., and Pullen, A. M., 1991, Superantigens: Mechanism of T-cell stimulation and role in immune responses, Ann. Rev. Immunol., 9:745.

Janeway Jr., C. A., Conrad, P. J., Tite, J. P., Jones, B., and Murphy, D. B., 1983, Efficiency of antigen presentation differs in mice differing at the Mls locus, Nature, 306:80.

Janeway Jr., C. A, Yagi, J., Conrad, P. J., Katz, M. E., Jones, B., Vroegop, S., and Buxser, S., 1989, T-cell responses to Mls and to bacterial proteins that mimic its behavior, Immunol. Rev., 107:61.

Kotb, M., Courtney, H. S., Dale, J. B., and Beachey, E. H., 1989, Cellular and biochemical responses of human T lyphocytes stimulated with Streptococcal M proteins, J. Immunol., 142:966.

Krzych, U., Fowler, A. V., Miller, A., and Sercarz, E., 1982, Repertoires of T cells directed against a large protein antigen, ß galactosidase. I. Helper cells have a more restricted specificity repertoire than proliferative cells, J. Immunol., 128:1529.

Liu, H., Lampe, M. A., Iregui, M. V., and Cantor H., 1991a, Conventional antigen and superantigen may be coupled to distinct and cooperative T-cell activation pathways, Proc. Natl. Acad. Sci. USA, 88:8705.

Liu, H., Lampe, M. A., Iregui, M. V., and Cantor, H., 1991b, Clonal unresponsiveness resulting from an interaction between SEB and T-cells expressing alternative V_β elements, Submitted for publication.

Marrack, P., and Kappler, J. W., 1990, The Staphylococcal Enterotoxins and their relatives, Science, 248:705.

Marrack, P., Blackman, M., Kushnir, E., and Kappler, J., 1990, The toxicity of staphylococcal enterotoxin B in mice mediated by T cells, J. Exp. Med., 171:455.

Mercep, M., Weissman, A. M., Frank, S. J., Klausner, R. D., and Ashwell, J. D., 1989, Activation-driven programmed cell death and T cell receptor zeta eta expression, Science, 246:1162.

Misfeldt, M. L., Legaard, P. K., Howell, S. E., Fornella, M. H., and LeGrand, R. D., 1990, Induction of Interleukin-1 from murin peritoneal macrophages by Pseudomonas aeruginosa, Inf. Imm., 58:978.

O'Hehir, R. E., and Lamb, J. R., 1990, Induction of specific clonal anergy in human T lymphocytes by Staphylococcus aureus enterotoxins, Proc. Natl. Acad. Sci. USA, 87:8884.

O'Rourke, A. M., Mescher, M. F., and Webb, S. R., 1990, Activation of polyphosphoinositide hydrolysis in T cells by H-2 alloantigen but not MLS determinants, Science, 249:171.

Patarca, R., Wei, F. -Y., Iregui, M. V., and Cantor, H., 1991, Differential induction of interferon- gene expression after activation of CD4$^+$ T cells by conventional antigen and Mls superantigen, Proc. Natl. Acad. Sci. USA, 88:2736.

Peterson, L. B., Wilner, G. D., and Thomas, D. W., 1983, Proliferating and helper T lymphocytes display distinct fine specificities in response to human fibrinopeptide B, J. Immunol., 130:2542.

Pinto, M., Torten, M., and Birnbaum, S. C., 1978, Suppression of the in vivo humoral and cellular immune response by Staphylococcal enterotoxin B, Transplantation, 25:320.

Rellahan, B. L., Jones, L. A., Kruisbeek, A. M., Fry, A. M., and Matis, L. A., 1990, In vivo induction of anergy in peripheral $V_\beta 8^+$ T cells by staphylococcal enterotoxin B, J. Exp. Med., 172:1091.

Schwartz, R., 1990, A cell culture model for T lymphocyte clonal anergy, Science, 248:1349.

Sussman, J. J., Mercep, M., Saito, T., Germain, R. N., Bonvini, E., and Ashwell, J. D., 1988, Dissociation of phosphoinositide hydrolysis and increased in intracellular Ca2+ from the biological response of a T cell hybridoma, Nature, 334:625.

Tite, J. P., Foellmer, H. G., Madri, J. A., and Janeway, Jr., C. A., 1987, Inverse Ir gene control of the antibody and T cell proliferative responses to human basement membrane collagen, J. Immunol., 139:2892.

White, J., Herman, A., Pullen, A. M., Kubo, R., Kappler, J. W., and Marrack, P., 1989, The V_β-specific superantigen Staphylococcal enterotoxin B: Stimulation of mature T cells and clonal deletion in neonatal mice, Cell, 56:27.

Whiteley, P. J., Poindexter, N. J., Landon, C., and Kapp, J. A., 1990, A peripheral mechanism preserves self-tolerance to a secreted protein in transgenic mice, J. Immunol., 145:1376.

Yagi, J., Baron, J., Buxser, S., and Janeway Jr., C. A., 1990, Bacterial proteins that mediate the association of a defined subset of T cell receptor: CD4 complexes with class II MHC, J. Immunol., 144:892.

19

MOLECULAR ANALYSIS OF THE INTERACTIONS BETWEEN STAPHYLOCOCCAL ENTEROTOXIN A SUPERANTIGEN AND THE HUMAN MHC CLASS II MOLECULES

L.N. Labrecque[*], J. Thidodeau[*], A. Herman[**], H. McGrath[*],
P. Marrack[**], J. Kappler[**], and R.-P. Sékaly[*]

[*]Laboratoire d'Immunologie, Institut de Recherches
Cliniques de Montréal, Montréal, Quebec, Canada H2W 1R7
[**]Department of Medicine, National Jewish Center
Denver, CO 80206

INTRODUCTION

Superantigens are a new class of antigen that stimulate a high proportion of T cells predominantly via the Vβ element of the T cell receptor (TcR) (1-3). Two groups of superantigens have been described; they include the mouse self-superantigen encoded by mouse mammary tumor viruses (4-6) and certain bacterial exotoxins (7-9).

The Staphylococcal enterotoxins (SEs) superantigens stimulate murine and human T cell in a Vβ specific manner. T cell stimulation by SEs requires direct binding of the superantigen to major histocompatibility complex (MHC) class II molecules (2,10-12). Unlike conventional antigens, SEs do not need to be processed to bind to MHC class II molecules and to stimulate T cells (12-14). Moreover, human DR molecules differ in their ability to present SEs to T lymphocytes. SEA and SEE bind well to DR1 but not to DRw53 (15). All HLA-DR molecules share a common α chain indicating that polymorphism in the β chain play an important role in SEs/MHC class II interactions.

To identify the amino acids in MHC class II molecules involved in the binding and presentation of SEA to T cells, we substituted in a reciprocal manner the residues that differ between the β chain of HLA-DR1 and HLA-DRw53. Our results demonstrate that the residue β81 of HLA-DR is critical for SEA binding to MHC class II molecules and for SEA presentation to T cells.

MATERIALS AND METHODS

Cell lines

3DT52.5.8 is a CD4 negative murine T cell hybridoma specific for D^d that expresses the TcR Vβ1 and Vβ8.1 genes (16).

Generation of mutant DR molecules

Site-specific mutagenesis of DRβ chain cDNA was performed using the PCR overlap extension technique as described by Ho *et al.* (17). The full-length PCR fragments were then digested with *Sac*I and *Stu*I and subcloned in a RSV.3 eukaryotic expression vector containing the wild-type DR1 or DRw53 cDNA and already digested with *Sac*I and *Stu*I to remove the wild-type 454 bp fragment. All of the mutants generated were confirmed by sequencing.

Chimeric DR1/DRw53 and DRw53/DR1 molecules that contain the first 146 a. a. of each DRβ chain cDNA were produced by exchanging the *Sac*I-*Stu*I fragment between the wild-type DR1 and DRw53 cDNAs.

Transfection of DRβ chain cDNAs

DNA transfections were performed using the calcium phosphate co-precipitation technique (18) and the MHC class II⁻ murine fibroblastic cell line DAP-3. DAP-3 cells expressing MHC class II molecules were obtained by two rounds of cell sorting with a FACSTAR PLUS (Becton Dickinson and Co.).

Cytofluorometric analysis of DAP-3 cells transfected with HLA-DR molecules

Cells were stained with either L243, a mouse anti-human class II antibody (5 μg/ml) which recognizes a non polymorphic determinant of DR, or 50D6 a mouse anti-human class II antibody which recognizes all DR except DR7 and DRw53, followed by fluorescein-labeled goat anti-mouse immunoglobulin. MHC class II expression was analyzed by flow cytometry using a FACScan (Becton Dickinson and Co.).

SEA stimulation of T cell hybridomas

Stimulation of 3DT52.5.8 by SEA was carried out as follows. 75 X 10³ T cells/well, were combined with 2 X 10⁴ DAP-3 fibroblasts expressing HLA-DR mutants and different concentrations of SEA in 96 well-plates, and incubated for 24 hours. The Interleukin-2 (IL-2) production was assayed by the ability of the culture supernatants to support the proliferation of the IL-2 dependent cell line CTLL.2 using the hexosaminidase colorimetric assay (19).

RESULTS AND DISCUSSION

The amino acid 81 of the β chain of HLA-DR is critical for SEA presentation to T cells

The staphylococcal exotoxin SEA has been shown to bind well to DR1 and poorly to DRw53 (15). Moreover, transfectants expressing DR1 but not DRw53 are efficient in the presentation of SEA to the murine T cell hybridoma 3DT52.5.8. To determine the residues on the MHC class II molecules which are important in the binding and presentation of SEA to T cells, we performed site-directed mutagenesis on the β chain of the HLA-DR molecules. Since the α chain of HLA-DR is monomorphic the differences in the binding and presentation of SEA involved the β chain of the MHC class II molecules. Protein sequence alignment between the first domain of the β chains encoded by the two alleles DR1 and DRw53 (Figure 1) showed that apart from the differences in polymorphic regions of the peptide binding groove, there are also differences in some of the exposed loops and in some of the conserved residues among most HLA-DR alleles (20). Using site-directed mutagenesis, we substituted the residue of DR1 by the corresponding residue in DRw53 and vice-versa.

We have previously shown that residue β81 of HLA-DR is critical for SEA binding (21). The substitution of the residue 81 (histidine) of the β1 domain of DR1 to a tyrosine (residue 81 of DRw53) abolishes the binding of SEA on DR1. Moreover, the

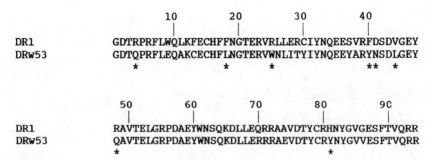

	10	20	30	40
DR1	GDTRPRFLWQLKFECHFFNGTERVRLLERCIYNQEESVRFDSDVGEY			
DRw53	GDTQPRFLEQAKCECHFLNGTERVWNLITYIYNQEEYARYNSDLGEY			
	*	* *	**	*

	50	60	70	80	90
DR1	RAVTELGRPDAEYWNSQKDLLEQRRAAVDTYCRHNYGVGESFTVQRR				
DRw53	QAVTELGRPDAEYWNSQKDLLERRRAEVDTYCRYNYGVVESFTVQRR				
	*			*	

Figure 1. Sequence alignment of the β1 domains of DR1 and DRw53 (20). Asterisks indicate the difference between DR1 and DRw53 outside of the polymorphic region of MHC class II molecules.

opposite mutation in DRw53 restores the binding of SEA to a level comparable to DR1. Mutations at other positions have no effect on SEA binding.

The HLA-DP molecule is able to present TSST-1 to T cells even if binding of TSST-1 can not be detected on this class II molecule (22). In order to determine if mutation β81 is sufficient to abolish or restore SEA presentation to T cells, we performed SEA stimulation of T cell hybridomas using the different class II mutants. The murine T cell hybridoma 3DT52.5.8 is stimulated when SEA is presented by fibroblasts expressing DR1 but not DRw53 (Figure 2a). In order to define which part of the DRβ chain is involved in the presentation of SEA, chimeric molecules between DR1 and DRw53 were used to stimulate T cells. The chimera DR1/DRw53 was as efficient as DR1 to stimulate 3DT52.5.8 but the chimera DRw53/DR1 was not able to present SEA to the T cell hybridoma (Figure 2a). These results clearly indicate that the differences observed in the first domain of the β chain of DR1 and DRw53 are responsible for the difference in SEA presentation. The mutants at position β81 of HLA-DR were then used to stimulate with SEA the 3DT52.5.8 hybridoma. The results (Figure 2b) show that the substitution H81Y in the β1 domain of DR1 abolishes the presentation of SEA to 3DT52.5.8. The small stimulation observed at high concentration of SEA (10 μg/ml) is similar to the one obtained using fibroblasts expressing DRw53 as stimulators. The substitution of tyrosine 81 of the β chain of DRw53 by an histidine is sufficient to get SEA stimulation of 3DT52.5.8 by DAP cells transfected with the DRw53 Y81H mutants (see Figure 2b). The class II molecule DRw53 Y81H is now as good as DR1 to present SEA to the T cell hybridoma 3DT52.5.8; the first concentration of SEA (1 ng/ml) that stimulates 3DT52.5.8 is the same for DR1 and DRw53 Y81H as shown by the dose-response curve (Figure 2b). These results show that residue 81 of the β1 domain of HLA-DR is critical for SEA binding and SEA stimulation of T cells.

Other groups have suggested that more than one SEA molecule can bind to MHC class II molecules (23, 24). Hence, it was surprising that mutation of only one residue in DR1 abolishes SEA binding. This effect is not due to a major change in the conformation of the molecule since this mutant is still able to present SEB (data not shown) and bind several monoclonal antibodies (Table 1). Moreover, the tyrosine is naturally found in the allele DRw53. Our results would suggest that only one binding site exists unless residue 81 overlaps two such sites. Another low affinity binding site for SEA might exist on class II molecules as suggested by the weak IL-2 production elicited by DRw53 at high toxin concentrations. It is possible that other residues in DR are involved in the binding site of SEA but residue 81 is the critical amino acid. This residue maybe important for the approach of the toxin to the MHC. Substitution of

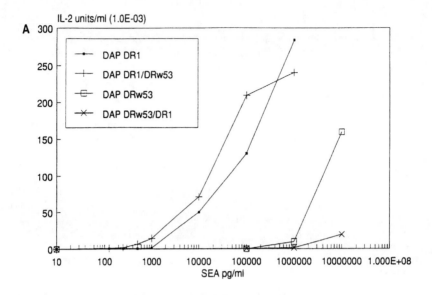

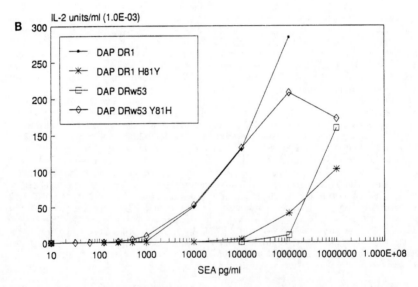

Figure 2. SEA stimulation of the mouse T cell hybridoma 3DT52.5.8. A) Dose response curve of 3DT52.5.8 to SEA presented by DAP DR1, DAP DRw53 and the chimeric molecules DR1/DRw53 and DRw53/DR1. B) Dose response curve of 3DT52.5.8 to SEA presented by DAP cells transfected with the DR1 H81Y mutant or the DRw53 Y81H. The stimulation of 3DT52.5.8 is indicated by Interleukin-2 production.

Table 1. Cytofluorometric analysis of DAP-3 cell lines expressing wild-type and mutant HLA-DR molecules

| CELL LINE[a] | MEAN FLUORESCENCE VALUE[b] | |
	L243	50D6
DAP DR1	21.10	9.26
DAP DR1/DRw53	11.80	5.89
DAP DR1 H81Y	40.56	36.71
DAP DRw53	21.60	2.30
DAP DRw53/DR1	14.15	1.68
DAP DRw53 Y81H	46.73	2.13

[a]DAP-3 cells were transfected with cDNAs encoding the DRα chain and different wild-type and mutant DRβ chains. Homogeneous populations of cells expressing the different class II molecules were obtained by aseptic cell sorting.

[b]Transfected cells were stained with DR specific monoclonal Abs L243 or 50D6 followed by a step with a goat anti-mouse Ig-FITC. Cells were analyzed on a FACscan. Mean fluorescence value were obtained on 10 000 live cells.

histidine by a tyrosine which is a bulky residue, would interfere directly or indirectly with the SEA class II interaction. Since histidine is a positively charged residue, it is likely that the nature of the charge plays an important role in the toxin-class II interaction. Experiments are under way to confirm both hypothesis. The differential affinity of SEA for I-E (10^{-6} M) (25) and DR (10^{-7} M) (10, 26) which both have histidine at position 81, suggests that other residues in the class II molecule affect the interaction with SEA.

Residue 81 has been suggested to be a TcR contact residue because of its side chain pointing upward the antigen binding groove (27). This could explain why DRw53 or DR1 H81Y not only affect the binding of SEA to class II but also the presentation to T cells unlike DP which does not bind TSST-1 but stimulates efficiently T cells (23).

The histidine β81 is highly conserved in human and murine MHC class II molecules. It is probably advantageous for the bacteria to produce toxins that bind to conserved residues of MHC class II molecules; this allows the bacteria to stimulate the immune system of all the individuals despite the strong polymorphism in MHC. Our preliminary experiments indicate that the binding sites of other toxins map to other highly conserved residues.

ACKNOWLEDGEMENTS

We thank Claude Cantin for flow cytometry expertise. This work was supported by grants to R.-P.S. from the Medical Research Council of Canada and the National Cancer Institute of Canada. N.L. is a student fellow of the Medical Research Council of Canada and J.T. is a postdoctoral fellow of the Fond de la Recherche en Santé du Québec.

REFERENCES

1. Janeway, C.A., Jr, Yagi, J., Conrad, P., Katz, M., Vroegop, S. and Buxser, S. T cell responses to Mls and to bacterial proteins that mimic its behavior. Immunol. Rev. 107:61-88, (1989).

2. White, J., Herman, A., Pullen, A.M., Kubo, R., Kappler, J. and Marrack, P. The Vβ-specific superantigen staphylococcal enterotoxin B: Stimulation of mature T cells and clonal deletion in neonatal mice. Cell 56:27-35, (1989).

3. Herman, A., Kappler, J.W., Marrack, P. and Pullen, A.M. Superantigens: Mechanism of T-cell stimulation and role in immune responses. Ann. Rev. Immunol. 9:745-772, (1991).

4. Woodland, D.L., Happ, M.P., Gollob, K.A. and Palmer, E. An endogeneous retrovirus mediating deletion of αβ T cells. Nature 349:529-530, (1991).

5. Acha-Orbea, H., Shakhov, A.N., Scarpellino, L., Kolb, E., Muller, V., Vessaz-Shaw, A., Fuchs, R., Blochlinger, K., Rollini, P., Billotte, J., Sarafidou, M., MacDonald, H.R. and Diggelman, H. Clonal deletion of Vβ14-bearing T cells in mice transgenic for mammary tumour virus. Nature 350:207-211, (1991).

6. Choi, Y., Kappler, J.W. and Marrack, P. A superantigen encoded in the open reading frame of the 3' long terminal repeat of mouse mammary tumour viruses. Nature 350:203-207, (1991).

7. Marrack, P. and Kappler, J.W. The staphylococcal enterotoxins and their relatives. Science 248:705-711, (1990).

8. Langford, M.P., Stanton, G.J. and Johnson, H.M. Biological effects of staphylococcal enterotoxin A on human peripheral lymphocytes. Infect. Immunol. 22:62-68, (1978).

9. Paevy, D.L., Alder, W.H. and Smith, R.T. The mitogenic effects of endotoxin and Staphylococcal enterotoxin B on mouse spleen cells and human peripheral lymphocytes. J. Immunol. 105:1453-1458, (1970).

10. Fraser, J.D. High affinity binding of staphylococcal enterotoxins A and B to HLA-DR. Nature 339:221-223, (1989).

11. Fleischer, B., Schrezenmeier, H. and Conradt, P. T lymphocyte activation by staphylococcal enterotoxins: role of class II molecules and T cell surface structures. Cell. Immunol. 120:92-101, (1989).

12. Fleischer, B. and Schrezenmeier, H. T cell stimulation by staphylococcal enterotoxins. Clonally variable response and requirements for malor histocompatibility complex class II molecules on accessory or target cells. J. Exp. Med. 167:1697-1707, (1988).

13. Yagi, J., Buxser, S. and Janeway, C.A., Jr. Bacterial proteins that mediate the association of a defined subset of T cell receptor:CD4 complexes with class II MHC. J. Immunol. 144:892-901.

14. Carlsson, R., Fisher, H. and Sjogren, H.O. Binding of staphylococcal enterotoxin A to accessory cells is a requirement for its ability to activate human T cells. J. Immunol. 140:2484-2488, (1988).

15. Herman, A., Croteau, G., Sékaly, R.-P., Kappler, J. and Marrack, P.   HLA-DR alleles differ in their ability to present staphylococcal enterotoxins to T cells. J. Exp. Med. 172:709-717, (1990).

16. Greenstein, J.L., Kappler, J., Marrack, P. and Burakoff, S.J. The role of L3T4 in recognition of Ia by a cytotoxic H-2D^d-specific T cell hybridoma. J. Exp. Med. 159:1213-1224, (1984).

17. Ho, S.N., Hunt, H.D., Horton, R.M., Pullen, J.K. and Pease, L.R.   Site-directed mutagenesis by overlap extension using the polymerase chain reaction.   Gene 77:51-59, (1989).

18. Graham, F.L. and Van der Eb, A.J. A new tchnique for the assay of infectivity of human adenovirus 5 DNA. Virology 52:456-467, (1973).

19. Landegren, U.   Measurement of cell numbers by means of the endogeneous enzyme hexosaminidase.   Applications to detection of lymphokines and cell surface antigens. J. Immunol. Methods 67:379-388, (1984).

20. Marsh, S.G.E. and Bodmer J.G.   HLA-DR and -DQ Epitopes and monoclonal antibody specificity. Immunology Today 10:305-312, 1989.

21. Herman, A., Labrecque, N., Thibodeau, J., Marrack, P., Kappler, J.W. and Sékaly, R.-P.   Identification of the staphylococcal enterotoxin A superantigen binding site in the β1 domain of the human histocompatibility Antigen HLA-DR. Proc. Natl. Acad. Sci. USA 88:9954-9958, (1991).

22. Scholl, P.R., Diez, A., Karr, R., Sékaly, R.-P., Trowsdale, J. and Geha, R.S. Effect of isotypes and allelic polymorphism on the binding of staphylococcal exotoxins to MHC class II molecules. J. Immunol. 144:226-230, (1990).

23. Pontzer, C.H., Russell, J.K. and Johnson, H.M.   Structural basis for differential binding of staphylococcal enterotoxin A and toxic shock syndrome toxin 1 to class II major histocompatibility molecules.   Proc. Natl. Acad. Sci. USA 88:125-128. (1991).

24. Mollick, J.A., Chintagumpala, M., Cook, R.G. and Rich, R.R.   Staphylococcal exotoxin activation of T cells. Role of exotoxin-MHC class II binding affinity and class II isotype. J. Immunol. 146:463-468, (1991).

25. Lee, J.M. and Watts, T.H. Binding of staphylococcal enterotoxin A to purified murine MHC class II molecules in supported lipid bilayers.   J. Immunol. 145:3360-3366, (1990).

26. Fisher, H., Dohlsten, M., Lindvall, M., Sjogren, H.O. and Carlsson, R. Binding of staphylococcal enterotoxin A to HLA-DR on B cell lines.   J. Immunol. 142:3151-3157, (1989).

27. Brown, J.H., Jardetzky, T., Saper, M.A., Samraoui, B., Bjorkman, P.J. and Wiley, D.C. A hypothetical model of the foreign antigen binding site of class II histocompatibility molecules. Nature 332:845-850, (1988).

20

IMMUNOSUPPRESSION BY MHC CLASS II BLOCKADE

Luciano Adorini

Preclinical Research
Sandoz Pharma Ltd.
CH-4002 Basel, Switzerland

INTRODUCTION

Antigen processing and presentation is a complex set of events leading to the activation of specific T cells by a binary ligand formed by antigenic peptides bound to major histocompatibility complex (MHC)-encoded molecules (1, 2).

Over the past few years considerable progress has been made in understanding antigen processing and presentation (3, 4). The essential steps involve intracellular antigen proteolysis in antigen-presenting cells (APC), association of peptides to MHC molecules and expression on the APC surface of the peptide-MHC complexes. Peptides bound to class I or class II MHC molecules on the surface of APC serve as ligands for the T cell receptors (TCR) of CD8$^+$ or CD4$^+$ cells, respectively. It is also becoming clear that the capacity of T cells to recognize self antigens is stringently controlled by multiple mechanisms, both thymic and peripheral, preventing inappropriate T cell activation potentially leading to autoimmune diseases (5).

Autoimmune diseases result from the activation of self-reactive T cells induced by autoantigens or by foreign antigens cross-reactive with an autoantigen. A striking characteristic of autoimmune diseases is the increased frequency of certain HLA alleles in affected individuals (6). Moreover, as demonstrated for example in rheumatoid arthritis and insulin-dependent diabetes mellitus, class II alleles positively associated with autoimmune diseases share amino acid residues in the hypervariable HLA regions involved in peptide binding (7). Therefore, it is likely that disease-associated HLA class II molecules have the capacity to bind the autoantigen and present it to T cells, thereby inducing and maintaining, under appropriate environmental conditions, the autoimmune disease. Since these diseases are almost invariably chronic and often very severe, they require effective long-term treatment.

The available drugs for the treatment of autoimmune diseases have not been developed with a clear understanding of the *primum movens* in the disease process, that is the activa-

T Lymphocytes: Structure, Function, Choices, Edited by F. Celada
and B. Pernis, Plenum Press, New York, 1992

tion of autoreactive T cells, and they are inadequate because of limited efficacy, modest if any selectivity, and considerable toxicity.

A new generation of immunosuppressive agents potentially useful in the prevention and/or treatment of autoimmunity is being fostered by recent progress in understanding antigen presentation to T cells, suggesting a number of suitable attack points for selective immunointervention (8).

The activation of potentially autoreactive T cells could be prevented by interfering with MHC, TCR, CD4 or other accessory molecules. All these target molecules could be interfered with in different ways, but two approaches are most commonly used. The first one relies on administration of monoclonal antibodies specific for molecules involved in antigen recognition and/or T cell activation. The second type of approach utilizes synthetic peptides designed to passively block MHC class II-restricted activation of autoreactive T cells (9), or to actively induce regulatory CD8$^+$ T cells able to suppress disease-inducing CD4$^+$ T cells (10).

Thus, the antigen presenting function of MHC class II molecules could be inhibited by anti-MHC antibodies, or by blocking the MHC binding site with a competitor ligand. Similarly, the T cell receptor could be targeted by specific antibodies against TCR variable regions or by antagonists binding to the TCR but inducing anergy rather than cell activation. Moreover, the interaction between CD4 and MHC class II could be inhibited by antibodies directed against either structures or by compounds preventing their interaction.

Administration of synthetic peptides represents a particularly attractive mode of selective immunointervention since peptides can be appropriately designed and, in general, they should be devoid of non-specific toxic effects. As alluded to before, immunotherapy by peptides could be distinguished in two major categories: i) passive therapies aimed at blocking T cell activation by blocking the MHC binding site to any antigenic peptide, including autoantigens, and ii) active therapies aimed at inducing or enhancing regulatory T cells able to control the disease, or at tolerizing the autoreactive T cells. This review will focus on immunosuppression by MHC class II-blocking peptides.

IMMUNOSUPPRESSION BY MHC-BINDING PEPTIDES

Immunosuppression by MHC binding peptides has been tested in a variety of experimental systems, by examining their effect on T cell activation, *in vivo*, either in the course of responses to conventional antigens (11, 12) or in autoimmune disease models (13-15).

These experimental systems could be divided in two groups, based on the presence of structural homology between the peptides used to inhibit and to induce T cell responses. Accordingly, as outlined in Table 1, inhibition of T cell activation by injection of class II-binding peptides could reflect different immunosuppressive mechanisms.

In situations when the inhibitor peptide is structurally closely related to the antigenic peptide, antigen-specific rather than MHC-specific mechanisms may be involved in the inhibition of T cell activation by administration of class II-binding peptides. These include induction of T cell tolerance (5), induction of suppressor T cells (10), or production by T cells of different lymphokines (16), such as IL-10, inhibiting Th1-type T cell clones. In addition, it is possible that homology between antigen and inhibitor could result in antagonism at the T cell receptor level, as suggested by prevention of experimental allergic encephalomyelitis (EAE) upon administration of an analogue of the encephalitogenic peptide (17).

Table 1. Mechanisms leading to inhibition of T cell activation by administration of class II-binding peptides

Homologous to antigen	Non-homologous to antigen

- T cell tolerance
- T cell anergy
- Induction of suppressor
  T cells
- Production of a different
  lymphokine profile
- MHC blockade
	- Clonal dominance
	  (if immunogenic)
	
	- MHC blockade

In mice of $H-2^u$ haplotype the acetylated N-terminal peptide (amino acid residues 1-9) of myelin basic protein is able to induce encephalitogenic T cells (18). Using this experimental model, several authors have demonstrated that injection of encephalitogenic peptides together with non pathogenic analogues can prevent the clinical development of EAE. Disease prevention has been related to competition for binding of encephalitogenic peptides to $I-A^u$ molecules (13, 14), with the consequent failure to activate encephalitogenic T cells. The evidence for this mechanism was based on the observation that peptides able to prevent EAE induction bind to $I-A^u$ better than encephalitogenic peptides, both in assays using purified class II molecules (13) and in competition for antigen presentation *in vitro* (14). However, since inhibitor and encephalitogenic peptides were highly homologous, induction of T cell tolerance or of suppressor T cells by the competitor peptide could also account for disease prevention (19).

If inhibitory and antigenic peptides are non homologous, but the inhibitory peptide is immunogenic, inhibition of the T cell response to antigen could result from MHC blockade or from clonal dominance induced by the inhibitory peptide. This situation is exemplified by the study of Lamont et al. (15), using a peptide derived from the murine proteolipid protein (PLP 139-151) to induce EAE in $H-2^s$ mice. Coadministration of this encephalitogenic peptide together with an unrelated peptide binding to $I-A^s$ prevents EAE induction (15). Since in this case encephalitogenic and competitor peptides do not share sequence homology, induction of T cell anergy or suppression by the competitor peptide does not appear to be the mechanism of disease prevention, and indeed this is most likely accounted for by MHC blockade. However, the inhibitory peptides used were themselves immunogenic, leaving open the possibility for a mechanism of clonal dominance in preventing induction of encephalitogenic T cells.

Collectively, these results indicate that several mechanisms, in addition to MHC blockade, may be responsible for in vivo inhibition of T cell activation by administration of MHC class II-binding peptides.

Our previous experiments have utilized a non immunogenic class II-binding peptide unrelated, in most cases, to the antigen, strongly suggesting MHC blockade as the cause for inhibition of T cell activation (11, 12). The peptide used in several experiments was the non-immunogenic self peptide mouse lysozyme (ML) 46-62, binding strongly to I-A^k molecules, but non binding to I-E^k or I-E^d molecules (11, 20). Previous experiments have demonstrated that injection of this peptide competitor inhibits in vivo T cell activation by I-A^k-binding antigenic peptides (11). Since the inhibition of T cell priming induced by a non-immunogenic competitor was specific for the class II molecule binding the competitor peptide, it could be inferred that the underlying in vivo mechanism was inhibition of antigen presentation by competitive binding to the class II molecule restricting the T cell response (9, 11, 12). Data reviewed in the following section provide now direct evidence for this mechanism.

DIRECT EVIDENCE FOR MHC BLOCKADE IN VIVO

To directly demonstrate MHC blockade, first we established an experimental model to detect complexes of naturally processed antigenic peptides generated in vivo and expressed by MHC class II molecules on the surface of APC. This was accomplished by immunizing mice with HEL in CFA and assessing the expression of antigenic complexes in APC draining the injection site by their ability to activate HEL peptide-specific, class II-restricted T cell hybridomas. The basic experimental design relied on the hypothesis that lymph node cells from HEL-primed mice would include antigen-presenting cells (APC) expressing on the cell surface detectable complexes of naturally processed HEL peptides bound to class II MHC molecules. To test this hypothesis C3H and DBA/2 mice were immunized with HEL-CFA and eight days later draining lymph node cells were irradiated and incubated, in the absence of added antigen, with the T cell hybridoma 3A9, recognizing the HEL peptide 46-61 together with I-A^k molecules, or with the T cell hybridoma 1H11.3, recognizing the HEL peptide 108-116 together with I-E^d molecules. Lymph node cells from HEL-primed C3H mice induce IL-2 production by 3A9 but not by 1H11.3 T cells. Conversely, lymph node cells from HEL-primed DBA/2 mice induce IL-2 production by 1H11.3 but not by 3A9 T cells. Lymph node cells from CFA-primed mice of either strain fail to induce activation of HEL-specific T cell hybridomas. Similarly, irradiated lymph node cells from HEL-primed mice, cultured alone, fail to produce detectable amounts of IL-2. These results show that lymph node cells from HEL-primed mice contain APC expressing complexes of HEL peptides and class II MHC molecules recognized by the appropriate T cell hybridomas.

To confirm that the antigen presenting activity of lymph node cells from HEL-primed mice was indeed MHC class II-restricted, monoclonal anti-I-A and anti-I-E antibodies were added to cultures of lymph node cells and T cell hybridomas. Anti-I-A, but not anti-I-E, antibodies abrogate antigen presentation by lymph node cells to the I-A^k-restricted hybridoma 1C5.1. Conversely, anti-I-E, but not anti-I-A, antibodies completely inhibit activation by lymph node cells from HEL-primed mice of the I-E^d-restricted hybridoma 1H11.3. We also assessed whether antigenic complexes are formed *in vivo*, rather than by carry-over of HEL, processed *in vitro* during the culture period. The experiment was carried out by incubating lymph node cells from HEL-primed mice with the lysosomotropic agent chloroquine, which failed to interfere with their antigen presenting capacity, indicating the presence of preformed complexes between antigenic peptides and class II molecules. (Guéry et al., manuscript in preparation).

Having established that detectable antigenic complexes are formed *in vivo* between class II MHC molecules and peptides derived from HEL processing, we examined the effect of administering a peptide competitor unrelated to the antigenic peptide on the formation of antigenic complexes. C3H mice were immunized with HEL and the HEL peptide 64-77, non binding to I-A^k, in CFA or with an emulsion containing HEL and the I-A^k-binding peptide HEL 112-129. Eight days later irradiated lymph node cells were tested for their capacity to activate the I-A^k-restricted hybridoma 1C5.1 specific for the HEL peptide 46-61. Co-injection of 100-fold molar excess of peptide competitor inhibits very efficiently the I-A^k-restricted antigen-presenting activity. Therefore, these results show that injection of a peptide binding to a defined MHC class II molecule selectively inhibits the *in vivo* formation of antigenic complexes by peptides binding to the blocked class II molecule. In addition, lymph node APC could present *in vitro* exogenous HEL to T cell hybridomas, even when mice were co-injected with the I-A^k-binding competitor peptide. This result indicates that APC from competitor-injected mice can rapidly regain, *in vitro*, full antigen-presenting capacity due to newly synthesized MHC class II molecules.

INHIBITION OF IN VIVO ANTIBODY RESPONSES BY MHC BLOCKADE

Since MHC blockade occurs *in vivo*, this could represent a promising approach to prevent and possibly also to treat autoimmune diseases. Selective MHC blockade can be induced by administering peptide competitors either in a depot (11, 12) or in soluble form (20). Moreover, exogenous competitors can inhibit, *in vitro*, equally well presentation of exogenous or endogenous antigens, the latter likely the most relevant in the induction of autoreactive T cells leading to HLA-associated autoimmune diseases (21).

However, several points need still to be addressed to evaluate the practical feasibility of this form of immunointervention. For example, thus far, MHC blockade has only been evaluated by its effect on T cell activation. It is not known whether *in vivo* competition for antigen presentation can also prevent induction of T cell-mediated antibody responses.

We have examined the effect of MHC class II blockade on the antibody response to T cell-dependent antigens and found that administration of MHC binding peptides profoundly inhibits the antibody response to antigens presented by the blocked MHC molecule. This inhibition is selective for the MHC class II molecules to which the competitor peptide binds and its extent depends on the molar ratio between antigenic and competitor peptides.

The self mouse lysozyme peptide corresponding to residues 46-62 (ML 46-62) binds to I-A^k molecules and it selectively inhibits, when coinjected with antigen, priming of I-A^k-restricted, antigen-specific T cells (11). Administration of ML 46-62 inhibits also antibody responses induced by HEL-specific, I-A^k-restricted helper T cells in B10.A(4R) mice. These mice, expressing only I-A^k as MHC class II molecule, were injected with the native HEL molecule, and the anti-HEL antibody response was measured by quantitative ELISA at different times after priming. Therefore, in this case, the T cell-dependent antibody response to HEL could only be mediated by I-A^k-restricted helper T cells, specific for the different immunodominant HEL peptides binding to I-A^k molecules, generated by HEL processing (12, 22). At least three immunodominant HEL determinants are recognized by I-A^k-restricted T cells. Two major ones are included in the HEL sequences 46-61 and 116-129, whereas a minor one is located in the HEL region 25-43 (12, 22). Since blocking the antigen-presenting function of the I-A^k molecule results in inhibition of the entire anti-HEL antibody response, T cell responses to all these determinants are apparently blocked, consistent with *in vivo* MHC blockade. As expected from its binding specificity, ML 46-62 does not inhibit the anti-HEL antibody response in mice of H-2^d haplotype.

The notion that MHC blockade is the mechanism leading to inhibition of antibody response by administration of ML 46-62 is also supported by the observation that this I-A^k-binding peptide can strongly inhibit, in B10.A(4R) mice, the antibody response to ribonuclease A, a protein antigen totally unrelated to the MHC blocker (Guéry et al., manuscript in preparation).

Inhibition of antibody responses by MHC blockade may have practical implications, since in some autoimmune diseases, e.g., in myasthenia gravis, pathogenic effects are primarily mediated by autoantibodies, rather than by autoreactive T cells.

CONCLUSIONS

MHC class II-binding peptides administered *in vivo* can be effective and selective inhibitors of T cell activation. We have demonstrated that injection of a class II-binding peptide can induce MHC blockade *in vivo*, resulting in inhibition of T cell activation and of T cell-dependent antibody responses. MHC blockade modulates T cell activation by interfering with the binding of antigenic peptides to class II molecules and this represents a promising approach to induce selective immunosuppression in HLA-associated autoimmune diseases (23, 24). The direct demonstration of MHC blockade *in vivo* obviously does not exclude the possible involvement of other peptide-mediated immunoregulatory mechanisms in experimental autoimmune models, particularly when antigen homologues are used as inhibitors. However, since a detailed characterization of human autoantigens is not yet available, antigen-specific immunosuppressive strategies are not yet feasible. Conversely, MHC blockade is already a validated approach to prevent experimental autoimmune diseases. It remains to be seen whether MHC blockers can also modify the course of human autoimmune diseases. Since MHC class II blockers are currently being developed for clinical testing by several groups, the answer should be forthcoming.

REFERENCES

1. Bjorkman, P.J., M.A. Saper, B. Samraoui, W.S. Bennett, J.L. Strominger, and D.C. Wiley. 1987. Structure of the human class I histocompatibiliy antigen, HLA-A2. *Nature* 329:506.
2. Brown, J.H., T. Jardetzky, M.A. Saper, B. Samraoui, P.J. Bjorkman, and D.C. Wiley. 1988. A hypothetical model of the foreign antigen binding site of class II histocompatibility molecules. *Nature* 332:845.
3. Moller, G. (ed.). 1988. Antigen processing. *Immunol. Rev.* 106:1-157.
4. Adorini, L. 1990. The presentation of antigen by MHC class II molecules. *Year Immunol.* 6:21.
5. Koshland, D.E., Jr. (ed.) 1990. Tolererance in the immune system. *Science* 248:1335.
6. Nepom G.T. 1988. Immunogenetics of HLA-associated diseases. *Concepts Immunopath.* 5:80.
7. Todd J.A., H. Acha-Orbea, J.I. Bell, N. Chao, Z. Fronek, C.O. Jacob, M. McDermott, A.A. Sinha, L. Timmermann, L. Steinman, and H.O. McDevitt. 1988. A molecular basis for MHC Class II-associated autoimmunity. *Science* 240:1003.
8. Adorini, L., V. Barnaba, C. Bona, F. Celada, A. Lanzavecchia, E. Sercarz, N. Suciu-Foca, and H. Wekerle. 1990. New perspectives on immunointervention in autoimmune diseases. *Immunol. Today* 11:383.
9. Adorini, L., and Z.A. Nagy. 1990. Peptide competition for antigen presentation. *Immunol. Today* 11:21.
10. Janeway, C. 1989. Immunotherapy by peptides. *Nature* 341:482.
11. Adorini, L., S. Muller, F. Cardinaux, P.V. Lehmann, F. Falcioni, and Z.A. Nagy. 1988. In vivo competition between self peptides and foreign antigens in T cell activation. *Nature* 334:623.
12. Adorini, L., E. Appella, G. Doria, and Z.A. Nagy. 1988. Mechanisms influencing the immunodominance of T cell determinants. *J. Exp. Med.* 168:2091.
13. Wraith D.C., D.E. Smilek, D.J. Mitchell, L. Steinman, and H.O. McDevitt. 1989. Antigen recognition in autoimmune encephalomyelitis and the potential for peptide-mediated immunotherapy. *Cell* 59:247.
14. Sakai, K., S.S. Zamvil, D.J. Mitchell, S. Hodgkinson, J.B. Rothbard, and L. Steinman. 1989. Prevention of experimental encephalomyelitis with peptides that block interaction of T cells with major histocompatibility complex protein. *Proc. Natl. Acad. Sci. USA* 86:9470.
15. Lamont A.G., A. Sette, R. Fujinami, S.M. Colon, C. Miles, and H.M. Grey. 1990. Inhibition of experimental autoimmune encephalomyelitis induction in SJL/J mice by using a peptide with high affinity for I-A^s molecules. *J. Immunol.* 145:1687.
16. Evavold, B.D., and P.M. Allen. 1991. Separation of IL-4 production from Th cell proliferation by an altered T cell receptor ligand. *Science* 252:1308.

16. Evavold, B.D., and P.M. Allen. 1991. Separation of IL-4 production from Th cell proliferation by an altered T cell receptor ligand. *Science* 252:1308.

17. Smilek, D.E., D.C. Wright, S. Hodgkinson, S. Dwivedy, L. Steinman, and H.O. McDevitt. A single amino acid change in a myelin basic protein peptide confers the capacity to prevent rather than induce experimental autoimmune encephalomyelitis. *Proc. natl. Acad Sci USA*, in press.

18. Zamvil, S.S., D.J. Mitchell, N.E. Lee, A.C. Moore, K. Kitamura, L. Steinman, and J.B. Rothbard. 1986. T cell epitope of the autoantigen myelin basic protein that induces encephalomyelitis. *Nature* 324:258.

19. Smilek, D.E., B.L. Lock, and H.O. McDevitt. 1990. Antigenic recognition and peptide-mediated immunotherapy in autoimmune disease. *Immunol. Rev.* 118:37.

20. Muller, S., L. Adorini, A. Juretic, and Z.A. Nagy. 1990. Selective in vivo inhibition of T cell activation by class II MHC-binding peptides administered in soluble form. *J. Immunol.* 145:4006.

21. Adorini, L., Moreno, J., Momburg, F., Hammerling, G.J., Guery, J-C., Valli, A., and Fuchs S. 1991. Exogenous peptides compete for the presentation of endogenous antigens to major histocompatibility class II-restricted T cells. *J. Exp. Med.* 174:0000.

22. Gammon, G., H.M. Geysen, R.J. Apple, E. Pickett, M. Palmer, A. Ametani, and E.E. Sercarz. 1991. T cell determinant structure: cores and determinant envelopes in three mouse majoe histocompatibility complex haplotypes. *J. Exp. Med.* 173:609.

23. Adorini, L. (ed.) 1990. The molecular basis of antigen presentation to T lymphocytes: novel possibilities for immunointervention. *Intern. Rev. Immunol.* 6:1-88.

24. Gefter, M.L. (ed.) 1991. Major Histocompatibility Complex and peptides in immunotherapy. *Sem. Immunol.* 3:193-255.

21

RECOGNITION OF HIV ANTIGENS

BY HUMAN T HELPER CELLS

Fabrizio Manca, Giuseppina Li Pira, Silvia Ratto
and Eleonora Molinari

Department of Immunology, University of Genoa
San Martino Hospital
16132 Genoa, Italy

INTRODUCTION

The interactions occurring between the human
immunodeficiency virus (HIV) and the human immune system are
not fully understood (1). The virus selectively infects
human CD4 positive T helper lymphocytes via the viral
external envelope glycoprotein gp120 (2,3) that binds human
CD4 molecules with high affinity and specificity (4-6).
Immune derangement follows infection (7) and accounts for
subsequent immunodeficiency that eventually results in the
clinically overt acquired immunodeficiency syndrome (AIDS)
(8,9).

IMMUNE RESPONSE TO HIV

The immune response to HIV antigens that can be
detected in infected individuals is apparently unable to
clear the infection, even though antibodies (10-14),
cytotoxic (15,16) and helper cells are present (17-20).
Therefore the protective mechanism(s) are not yet defined.
More information on protective mechanisms is highly
desirable for the development of new generation vaccines
(21,22) aimed at inducing immune responses possibly
differing in quantitative and qualitative terms from the
naturally occurring responses, that are poorly effective.

CENTRAL ROLE OF T HELPER CELLS

By analogy with other viral infections (23), it is
reasonable to predict that neutralizing antibodies (NA)
(24,25) and cytotoxic T lymphocytes (CTL) (23) represent
important effector mechanisms for eradication, or at least
for control of HIV infection. Both effector mechanisms imply
clonal expansion of NA producing B cells and of specific
CTL, that depends on intervention of T helper cells (26).

T Lymphocytes: Structure, Function, Choices, Edited by F. Celada
and B. Pernis, Plenum Press, New York, 1992

T helper cells, in fact, produce essential lymphokines upon antigen driven activation (26). Therefore up-stream in the immune cascade activation of virus specific T helper cells is a key event. For these reasons we decided to investigate the human T helper response and the T repertoire specific for HIV proteins.

T HELPER REPERTOIRE

Since HIV infection per se affects T helper lymphocytes (2,3) we started our investigation in non-HIV infected individuals, in order to avoid the interference of the immunosuppressive activity of HIV itself. By studying non infected-non immunized individuals we could examine the status of the "naive" repertoire of T helper cells specific for HIV (27). The possibility for such an investigation was suggested by previous reports (28,29) showing that by repeated in vitro priming and restimulation with antigen and antigen presenting cells (APC) it is possible to select the unfrequent precursors available in the naive repertoire and expand them to generate established T cell lines and clones. Such lines and clones were analyzed for fine specificity and for APC requirements.

BREADTH OF THE HIV SPECIFIC NAIVE T REPERTOIRE

T cell lines were generated from naive individuals by repeated in vitro priming with the envelope glycoprotein of HIV gp120. T lines were successfully generated from approximately 50% of the individuals examined. The established lines were tested for fine specificity by using a panel of 52 overlapping peptides (sixteen-mers with six aa overlap). All the lines were specific for only one or two of the peptides in the panel, indicating that the primary response to the antigenic protein focuses on a very limited region (27). Furthermore, each individual recognized as "immunodominant" a different region of gp120, suggesting that the MHC class II background affects recognition of different epitopes. Work to demonstrate that MHC class II actually dictates which are the immunodominant regions of gp120 is in progress. Since the majority of the gp120 sequence is not recognized under these experimental conditions, we tried to give the "immunorecessive" regions a chance to become immunogenic (27). A new set of T cell lines was generated by using the peptides for priming, either as a pool of 52 or as pools of 4 adjacent peptides (tetrapools). By using peptides for priming, the existence of a much broader repertoire was detected, suggesting that immunodominance of a peptide depends on both the MHC structure of the responder and on the molecular context an epitope is flanked by.

ROLE OF FLANKING SEQUENCES ON ANTIGENICITY

The concept of the molecular framework in which a T cell epitope is contained was originally proposed in the lysozyme system (30). We decided to examine this phenomenon

196

in greater detail in the human response to gp120 (31). In the naive individual used for most of the experiments the sequence corresponding to residues 236-251 of gp120 behaves as immunodominant. The same region has different aa residues in gp120 molecules from different strains, even though a six aa portion is conserved. In this system, therefore, it is possible to examine the immunogenicity of the conserved (consensus) region that is flanked by strain specific variable sequences. We showed that precursor T cells exist with specificity for the consensus region, but such T cells are triggered only when the consensus is flanked by appropriate sequences. This phenomenon can be attributed to the fact that flanking sequences may behave as "immunodominant" or as "immunorecessive" determinants, and therefore compete with the consensus determinant e.g. for MHC association and restricted presentation to the available naive repertoire. Alternatively they may dictate unique processing patterns within the APC.

T CELL REGULATION OF THE "QUALITY" OF B CELL RESPONSE

A central problem in the design of an effective vaccine aimed at inducing an appropriate T helper response is to trigger a set of T cell clones with predetermined fine specificities in order to drive also the antibody response towards the production of neutralizing antibodies. The possibility of directing the antibody response to conformational determinants on antigen molecules by selecting the fine specificity of T helper cells has been previously shown in a murine experimental system by using beta-galactosidase as antigen (32-34). With this premise we would like to propose that also in the HIV system the fine specificity of the antibody response could and should be regulated via T helper cells. It is therefore of paramount importance to learn all of the possible ways to control the "quality" of T helper cell response.

ROLE OF DIFFERENT ANTIGEN PRESENTING CELLS

In addition to the possible use of peptides (as such, as cocktails, as conjugates, etc.) to trigger selected T helper clones, optimization at the antigen presentation level may also provide useful information for vaccine design to induce response to poorly immunogenic peptides. Since antigen presentation experiments are hard, if not impossible to perform in vivo in humans, we took advantage of the available T cell lines and clones specific for gp120 (27) to define the optimal APC populations for secondary and for primary stimulation (35). In both cases populations of APC highly enriched in dendritic cells (DC) proved the most effective on a per cell basis. In particular DC were essential for induction of primary in vitro responses that resulted in the generation of established lines. This suggests that autologous irradiated blood mononuclear cells function as APC for primary immunization thanks to their content in DC. The other populations with APC function (monocytes and B cells) present in peripheral blood, in fact, do not provide effective primary presentation (35).

ENHANCED ANTIGEN UPTAKE (receptor mediated endocytosis)

Although primary immunization seems to rely mostly on DC, secondary stimulation also makes use of cells of the mononuclear lineage and of B cells. In this case enhanced uptake of antigen complexed with specific antibodies can be achieved by exploiting Fc receptors on these cells, as shown previously in other experimental systems in mice (36) and in humans (37). An additional way to facilitate antigen uptake by using functional receptors is to treat gp120 with neuraminidase. This enzyme cleaves terminal sialic acid molecules from the carbohydrate side chains of gp120 (a heavily glycosylated protein with sugars accounting for 50% of its molecular mass) (38). This exposes the underlying galactose residues that are promptly recognized by galactose receptors on monocytes and on DC, resulting in receptor mediated uptake of gp120 (39).

HUMAN T HELPER RESPONSE TO OTHER HIV PROTEINS

In addition to gp120 we have investigated the human naive repertoire specific for additional HIV proteins. Data are presently available on gp41 (40), the transmembrane component of the envelope glycoprotein that anchors gp120 non covalently to the virus surface, and on reverse transcriptase (p66) (41), the product of the pol gene of HIV. Lines have been generated by in vitro priming with gp41 peptides. Such lines generated from a large fraction of the individuals tested recognize different peptides, suggesting that also in this case MHC may dictate immunodominance. In addition, only a small percentage of these lines also recognize the whole protein fed into the system as gp160 (the covalent complex of gp41 plus gp120). Since gp41 is not available in pure form, but only in association with gp120, we are now trying to separate gp41 and gp120 on SDS gels that can be blotted on nitrocellulose, in order to obtain gp41 on a solid support that can be used effectively for T cell stimulation. This protocol has been successful with other antigens (42) for presentation to established lines and clones, but it is not known yet whether it also allows primary stimulation of naive T cells.
Reverse transcriptase specific T cell lines have been established from naive individuals with a high success rate (41). Since a panel of peptides is not yet available to our lab, we cannot define the fine specificity of the lines and of the clones to test for a possible individual MHC effect also in presentation of this antigen.

T CELL RESPONSE TO HIV ANTIGENS IN IMMUNIZED INDIVIDUALS

Humans can be immunized in vivo to HIV antigens upon infection (17-20,43-48) or upon deliberate administration of antigen for vaccination purposes (49,50). In both cases their T helper response has been studied and described in several reports. This type of information should be complemented with the data obtained from naive individuals, since in both immunized groups the immune response may be biased. In the case of infected individuals the persistence of the antigenic stimulus may select some clones and the presence of the CD4tropic virus may selectively delete some T cell specificities. In the case of immunized individuals the administration of whole antigen may severely select for T cell clones specific for immunodominant epitopes.

GP120 AS A CD4 BINDING REACTANT: INTERFERENCE WITH IMMUNE RESPONSE

gp120 may be released by HIV infected cells (51-53) and may absorb on CD4 positive cells (e.g. T helper cells, monocytes). This event has several important implications with respect to the immune response:
1. gp120 binding to CD4 molecules on T helper cells blocks antigen driven activation (54-56), that depends on integrity of CD4 as an accessory molecule.
2. gp120 binding to CD4 molecules on by-standing T helper cells drives ADCC effector cells on erroneous targets in the presence of gp120 specific antibodies (57,58).
3. gp120 binding to CD4 cells results in internalization and reexpression of gp120 peptides in association with MHC molecules that are recognized by class II restricted specific CTL that attack innocent targets (28,29,59).
4. gp120 binding to CD4 molecules on T helper cells (28) and on monocytes (60,61) results in enhanced uptake of the antigen and facilitates presentation to gp120 specific T helper cells.
Therefore it is evident that gp120 may exhibit immunoinhibitory activity , at least in *in vitro* experiments. If the same events also take place in vivo, additional mechanisms for immunodeficiency add up to the cytopathic effect of HIV for T helper cells.
It is also evident that gp120 displays a double nature, in that it can function both as an antigen and as an inhibitor for the immune system. These two activities can be dissociated by abolishing CD4 binding capacity either by denaturation or by complexing with soluble CD4 or with antibodies. In all of these instances antigenicity for specific T helper cells is retained (62).

QUALITY VS. QUANTITY OF THE IMMUNE RESPONSE

In spite of a multifaceted immune response to viral antigens in infected individuals, HIV infection progresses in most of the cases. Therefore one could conclude that the immune system is unable to cope with this virus and that only non-immunological approaches (pharmacology, molecular biology) may be effective for prevention or for treatment. Here we propose that the quality, rather than the quantity of the immune response may be inadequate and therefore manipulation of the T helper response may provide clues for the induction of a qualitatively appropriate response.

Acknowledgements

This work was financed by grants from the Italian Ministry of Health (AIDS Project, n. 5206.022) and from the National Research Council (CNR, Rome, n. 89.00913.72), and supported by the Royal Society (London) - CNR (Rome) agreement.
We are grateful to Franco Muzzì for valuable secretarial help.

REFERENCES

1. S. Koenig and A. S. Fauci, AIDS: Immunopathogenesis and immune response to the human immunodeficiency virus, in: "AIDS. Etiology, Diagnosis, Treatment, and Prevention. Second Edition", V. T. DeVita Jr., S. Hellman, S. A. Rosenberg, ed., J. B. Lippincott Company, Philadelphia (1990).

2. A. G. Dalgleish, P. C. L. Beverley, P. Clapham, D. Crawford, M. Greaves, and R. A. Weiss, The T4 (CD4) molecule is an essential component of the HTLVIII/LAV-1 receptor, Nature 312:767 (1984).

3. D. Klatzman, E. Champagne, S. Chamarets, D. Guetard, T. Hercend, J. C. Gluckman, and L. Montaignier, T lymphocyte T4 molecule behaves as the receptor for human retrovirus LAV, Nature 312:767 (1984).

4. L. A. Lasky, G. Nakamura, D. H. Smith, C. Fennie, C. Shimaski, E. Patzer, P. Berman, T. Gregory, and D. Capon, Delineation of a region of the human immunodeficiency virus type 1 gp120 glycoprotein critical for interaction with the CD4 receptor, Cell 50:975 (1987).

5. D. J. Capon and R. H. R. Ward, The CD4-gp120 interaction and AIDS pathogenesis, Annu. Rev. Immunol. 9:649 (1991).

6. T. Kieber-Emmons, B. A. Jameson, and W. J. W. Morrow, The gp120-CD4 interface: structural, immunological and pathological considerations, Biochim. et Biophysica Acta 989:281 (1989).

7. H. C. Lane, A. S. Fauci, Immunologic abnormalities in the acquired immunodeficiency syndrome, Annu. Rev. Immunol. 3:477 (1985).

8. M. S. Gottlieb, R. Schroff, H. M. Schanker, J. D. Weisman, P. T. Fan, R. A. Wolf, A. Saxon, Pneumocystis carinii pneumonia and mucosal candidias in previously healthy homosexual men: evidence for a new acquired cellular immunodeficiency, N. Engl. J. Med., 305:1425 (1981).

9. F. P. Siegal, C. Lopez, G. S. Hammer, A. E. Brown, S. J. Kornfeld, J. Gold, J. Hassett, S. Z. Hirschman, C. Cunningham, B. R. Adelsberg, Severe acquired immunodeficiency in homosexual males, manifested by chronic perianal ulcerative herpes simplex lesions, N. Engl. J. Med. 305:1439 (1981).

10. T. J. Matthews, A. J. Langlois, W. G. Robey, N. T. Chang, and R. C. Gallo, Restricted neutralization of divergent human T-lymphotropic virus type III isolates by antibodies to the major envelope

glycoprotein, Proc. Natl. Acad. Sci. USA 83:9709 (1986).

11. M. Robert-Guroff, J. M. Oleske, E. M. Connor, L. G. Epstein, and R. C. Gallo, Relationship between HTLV-III neutralizing antibody and clinical status of pediatric acquired immunodeficiency syndrome (AIDS) and AIDS related complex cases, Pediatr. Res. 21:547 (1987).

12. I. Wendler, U. Bienzele, G. Hunsmann, Neutralizing antibodies and the course of HIV-induced diseases, AIDS 3:157 (1987).

13. K. S. Steimer, G. V. Nest, D. Dina, P. J. Barr, P. A. Luciw, and E. T. Miller, Genetically engineered human immunodeficiency envelope glycoprotein gp120 produced in yeast is the target of neutralizing antibodies, Vaccine 877:236 (1987).

14. P. J. Klasse, and J. Blomberg, Patterns of antibodies to human immunodeficiency virus proteins in different subclasses of IgG, J. Infect. Dis. 156:1026 (1987).

15. B. D. Walker, S. Chakrabarti, B. Moss, T. J. Paradis, T. Flynn, A. G. Durno, R. S. Blumberg, J. C. Kaplan, M. S. Hirsch, and R. T. Schooley, HIV-specific cellular cytotoxic T lymphocytes in seropositive individuals, Nature 328:345 (1987).

16. D. S. Tyler, C. L. Nastala, S. D. Stanley, T. J. Matthews, H. K. Lyerly, D. P. Bolognesi, and K. J. Weinhold, GP120 specific cellular cytotoxicity in HIV-1 seropositive individuals. Evidence for circulating CD16+ effector cells armed in vivo with cytophilic antibody, J. Immunol. 142:1177 (1989).

17. R. D. Shrier, J. W. Ghann Jr, A. J. Langlois, K. Shriver, J. A. Nelson, and M. B. A. Oldstone, B- and T-lymphocyte responses to an immunodominant epitope of human immunodeficiency virus, J. Virol. 62:2531 (1988).

18. R. D. Shrier, J. W. Gnann Jr, R. Landes, C. Lockshin, D. Richman, A. McCutchan, C. Kennedy, M. B. A. Oldstone, and J. A. Nelson, T cell recognition of HIV synthetic peptides in a natural infection, J. Immunol. 142:1166 (1989).

19. J. V. Torseth, P. W. Berman, and T. C. Merigan, Recombinant HIV structural proteins detect specific cellular immunity in vitro in infected individuals, AIDS Res. Human Retroviruses 4:120 (1988).

20. T. Mathiesen, P. A. Broliden, J. Rosen, and B. Wahren, Mapping of IgG subclass and T-cell epitopes on HIV proteins by synthetic peptides, Immunology 67:453 (1989).

21. K. H. G. Mills, D. F. Nixon, and A. J. McMichaels, T-cells strategies in AIDS vaccines: MHC-restricted T-cell responses to HIV proteins, AIDS 3:S101 (1989).

22. D. P. Bolognesi, HIV antibodies and vaccine design, AIDS 3:S111 (1989).

23. J. M. Zarling, J. W. Eichberg, P. A. Moran, J. McClure, P. Sridhar, and H. Shiu-Lok, Proliferative and cytotoxic T cells to AIDS virus glycoproteins in chimpanzees immunized with a recombinant vaccinia virus expressing AIDS virus envelope glycoproteins, J. Immunol. 139:988 (1987).

24. M. Robert-Guroff, M. Brown, and R. C. Gallo, HTLV-III neutralizing antibodies in patients with AIDS and AIDS related complex, Nature 316:72 (1985).

25. R. A. Weiss, P. R. Clapham, R. Cheinsong-Popov, A. G. Dalgleish, C. A. Carne, I. V. Weller, and R. S. Tedder, Neutralization of human T-lymphotropic virus type III by sera of AIDS and AIDS-risk patients, Nature 316:69 (1985).

26. A. K. Abbas, M. E. Williams, H. J. Burstein, T.-L. Chang, P. Bossu, and A. H. Lichtman, Activation and functions of CD4+ T-Cell subsets, Immunol. Rev. 123 (1991).

27. F. Manca, J. Habeshaw, and A. Dalgleish, The naive repertoire of human T helper cells specific for gp120, the envelope glycoprotein of HIV, J. Immunol. 146:1964 (1991).

28. A. Lanzavecchia, E. Roosnek, T. Gregory, P. Berman, and S. Abrignani, T cells can present antigens such as HIV gp120 targeted to their own surface molecules, Nature 334:530 (1988).

29. R. F. Siciliano, T. Lawton, C. Knall, R. W. Karr, P. Berman, T. Gregory, and E. Reinherz, Analysis of host-virus interactions in AIDS with anti-gp120 T cell clones: effect of HIV sequence variation and a mechanism for CD4+ cell depletion, Cell 54:561 (1988).

30. G. Gammon, N. Shastri, J. Cogswell, S. Wilbur, S. Sadegh Nasseri, U. Krzych, A. Miller, and E. Sercarz, The choice of T-cell epitopes utilized on a protein antigen depends on multiple factors distant from, as well as the determinant site. Immunol. Rev. 98:53 (1987).

31. F. Manca, J. A. Habeshaw, A. G. Dalgleish, and E. E. Sercarz, Human T helper clones react differentially with strain specific variants of an immunodominant peptide of HIV gp120: role of flanking sequences in antigenicity, submitted.

32. F. Manca, A. Kunkl, D. Fenoglio, A. Fowler, E. Sercarz, and F. Celada, Constraints in T-B cooperation related to epitope topology on E. coly beta-galactosidase. The fine specificity of T cells dictates the fine specificity of antibodies directed to conformation dependent determinants, Eur. J. Immunol. 15:345 (1985).

33. F. Celada, and F. Manca, Specific T-B interaction in the response to conformation dependent determinants, in: " Synthetic Vaccines", R. Arnon, ed., CRC Press, Boca Raton (1987).

34. F. Manca, D. Fenoglio, A. Kunkl, C. Cambiaggi, G. Li Pira, and F. Celada, B cells on the podium: regulatory roles of surface and secreted immunoglobulins, Immunol. Today 9:300 (1988).

35. F. Manca, J. Habeshaw, and A. Dalgleish, Dendritic cells: the chief antigen presenting cells for in vitro induction of primary human CD4+ T cell lines specific for HIV envelope glycoprotein, submitted.

36. F. Manca, D. Fenoglio, A. Kunkl, C. Cambiaggi, M. Sasso, and F. Celada, Differential activation of T cell clones stimulated by macrophages exposed to antigen complexed with monoclonal antibodies. A possible influence of paratope specificity on the mode of antigen processing, J. Immunol. 140:1 (1988).

37. F. Manca, D. Fenoglio, G. Li Pira, A. Kunkl, and F. Celada, Effect of antigen antibody ratio on macrophages uptake, processing and presentation to T cells of antigen complexed with polyclonal antibodies, J. Exp. Med. 173:37 (1991).

38. C. K. Leonard, M. W. Spellman, L. Riddle, R. J. Harris, J. N. Thomas, T. J. Gregory, Assignment of intrachain disulfide bonds and characterization of potential glycosylation sites of the type I recombinant human immunodeficiency virus envelope glycoprotein (gp120) expressed in chinese hamster ovary cells, J. Biol. Chem. 265:10373 (1990).

39. F. Manca, Enhanced presentation of HIV asialo env glycoprotein to specific human T cells as evidence for a functional galactose receptor on human antigen presenting cells, submitted.

40. F. Manca, M. T. Valle, D. Fenoglio, A. Kunkl, Generation of human T cell lines specific for HIV gp41 peptides and gp41 protein from unprimed individuals, manuscript in preparation.

41. F. Manca, A. Kunkl, D. Fenoglio, M. T. Valle, T helper cells specific for HIV reverse transcriptase can be

selected in vitro from unprimed individuals, manuscript in preparation.

42. F. Manca, G. Rossi, M. Valle, S. Lantero, G. Damiani, G. Li Pira, and F. Celada, Limited clonal heterogeneity of antigen specific T cells focussing in the pleural space during mycobacterial infection, Infection and Immunity 59:503 (1991).

43. T. Mathiesen, P. A. Broliden, J. Rosen, and B. Wahren, Mapping of IgG subclass and T-cell epitopes on HIV proteins by synthetic peptides, Immunology 67:453 (1989).

44. C. M. Walker, K. S. Steimer, K. Rosenthal, and J. A. Levy, Identification of human immunodeficiency virus (HIV) envelope type-specific T helper cells in an HIV-infected individual. J. Clin. Invest. 82:2172 (1988),

45. J. Krowka, D. Stites, R. Debs, C. Larsen, J. Fedor, E. Brunette, and N. Duzgunes, Lymphocytes proliferative responses to soluble and liposome-conjugated envelope peptides of HIV-1, J. Immunol. 144:2535 (1990).

46. B. Wahren, L. Morfeldt-Mansson, G. Biberfeld, L. Moberg, A. Sonnerborg, P. Ljungman, A. Werner, R. Kurth, R. C. Gallo, and D. Bolognesi, Characteristics of the specific cell mediated immune response in human immunodeficiency virus infection, J. Virol. 61:2017 (1987).

47. P. M. Ahearne, T. J. Matthews, H. K. Lyerly, G. C. White, D. Bolognesi, and K. J. Weinhold, Cellular immune response to viral peptides in patients exposed to HIV, AIDS Res. Hum. Retrovir. 4:259 (1988).

48. B. Wahren, J. Rosen, E. Sandstrom, T. Mathiesen, S. Modrow, and H. Wigzell, HIV-1 peptides induce a proliferative response in lymphocytes from infected person, J. AIDS 4:448 (1989).

49. D. Zagury, J. Bernard, R. Cheynier, I. Desportes, R. Leonard, M. Fouchard, B. Reveil, D. Ittele, Z. Lurhuman, K. Mbayo, J. Wane, J. J. Salaun, B. Goussard, L. Dechazal, A. Burny, P. Nara, and R. C. Gallo, A group specific anamnestic immune reaction against HIV-1 induce by a candidate vaccine against AIDS, Nature 332:728 (1989).

50. S. Abrignani, D. Montagna, M. Jeannet, J. Wintsch, N. L. Higwood, J. R. Shuster, K. S. Steimer, A. Cruchaud, and T. Staehelin, Priming of CD4+ T cells specific for conserved regions of human immunodeficiency virus glycoprotein gp120 in human immunized with a recombinant envelope protein, Proc. Natl. Acad. Sci. USA 87:6136 (1990).

51. H. R. Gelderblom, H. Reupke, G. Ozel Pauli, Loss of envelope antigens of HTLV-III/LAV: a factor in AIDS pathology, Lancet 2:1016 (1985).

52. J. Schneider, P. Kaaden, T. D. Copeland, Shedding and interspecies type sero-reactivity of the envelope glycopolypeptide gp120 of the human immunodeficiency virus, J. Gen. Virol. 67:2533 (1986).

53. H. R. Gelderblom, G. Ozel Pauli, Morphogenesis and morphology of HIV structure-function relations, Arch. Virol. 106:1 (1989).

54. J. Habeshaw, A. G. Dalgleish, The relevance of HIV env/CD4 interactions to the pathogenesis of acquired immune deficiency syndrome, J. AIDS 2:457 (1989).

55. F. Manca, J. A. Habeshaw, A. G. Dalgleish, HIV envelope glycoprotein antigen specific T cell responses and soluble CD4, Lancet 335:811 (1990).

56. J. A. Habeshaw, A. G. Dalgleish, L. Bountiff, A. L. Newell, D. Wilks, L. C. Walker, and F. Manca, AIDS pathogenesis: HIV envelope and its interaction with cell proteins, Immunol. Today 11:418 (1990).

57. H. K. Lyerly, T. J. Matthews, A. Langlois, D. Bolognesi, K. Weinhold, Human T-cell lymphotropic virus IIB glycoprotein (gp120) bound to CD4 determinants on normal lymphocytes and expressed by infected cells serves as a target for immune attack, Proc. Natl. Acad. Sci. USA 84:4601 (1987).

58. D. S. Tyler, H. K. Lyerly, K. J. Weinhold, Minireview: anti-HIV ADCC, AIDS Res. Hum. Retrovir. 5:557 (1989).

59. R. Germain, Antigen processing and CD4+ T cell depletion in AIDS, Cell 54:441 (1988).

60. D. S. Finbloom, D. L. Hoover, and M. S. Meltzer, Binding of recombinant HIV coat protein gp120 to human monocytes, J. Immunol. 146:1316 (1991).

61. R. F. Siliciano, C. Knall, T. Lawton, P. Berman, T. Gregory, and E. L. Reinherz, Recognition of HIV glycoprotein gp120 by T cells. Role of monocytes CD4 in the presentation of gp120, J. Immunol. 142:1506 (1989).

62. F. Manca, L. Walker, A. Newell, F. Celada, J. A. Habeshaw, A. G. Dalgleish, Inhibitory activity of HIV envelope gp120 dominates over its antigenicity for human T cells, Clin. Exp. Immunol. In press.

22

PROLIFERATIVE RESPONSES TO THE V3 REGION OF HIV ENVELOPE
ARE ENHANCED FOLLOWING IMMUNIZATION WITH V3:Ty VIRUS-LIKE
PARTICLES IN MICE

Stephen J. Harris,[1] Andrew J.H. Gearing,[1] Guy T. Layton,[1]
Sally E. Adams,[1] Susan M. Kingsman,[2] and Alan J. Kingsman[1,2]

[1] British Bio-technology Ltd., Watlington Road, Cowley, Oxford, UK,
OX4 5LY

[2] Virus Molecular Biology Group, Department of Biochemistry, Oxford
University, South Parks Road, Oxford, UK, OX1 3QU

INTRODUCTION

T helper(Th)-lymphocytes play a vital role in the immune response to an invading
organism. The activation of CD4-positive Th lymphocytes in the production of cytokines
which enhance the function of cytotoxic T-lymphocytes (CTL), promote differentiation of
B-lymphocytes into antibody-producing plasma cells, and increase the frequency of
memory B-lymphocytes. It is likely that both of these arms of the immune response will
be involved in protection against the human immunodeficiency virus (HIV). A strong
anti-HIV CD8-positive activity is associated with increased long term survival (1).
Neutralizing antibodies directed against the HIV envelope have been described and
evidence presented which suggests a key protective role for these antibodies (2,3). An
HIV vaccine will, therefore, need to include epitopes recognized by Th lymphocytes.

We are currently developing a recombinant HIV vaccine. Fusion proteins comprising the
p1 protein of the yeast transposon Ty and the V3 loop sequence from the HIV isolate IIIB
assemble into 50nm virus-like particles (V3:Ty-VLPs) containing approximately 300
copies of the p1:fusion protein. The V3 region of HIV envelope was selected because it
contains both Th (4) and CTL (5) epitopes and also the principal antibody neutralizing
domain of HIV (1).

We have evaluated the relative efficacy of presenting this region in a particulate form by
measuring lymph node proliferative responses to V3 following immunization with
V3:Ty-VLPs, rgp120 and a V3-albumin conjugate (V3-alb). We have also investigated
the effect of adjuvant and immunization dose.

T Lymphocytes: Structure, Function, Choices, Edited by F. Celada
and B. Pernis, Plenum Press, New York, 1992

MATERIALS AND METHODS

V3 sequence: NCTRPNNNTRKRIRIQRGPGRAFVTIGKIGNMRQAHCNIS

V3 peptide: A 40mer corresponding to the above sequence and representing the V3 loop of HIV-1 isolate IIIB, clone HXB2 was obtained from Cambridge Research Biochemicals.

V3-alb: The peptide was conjugated to bovine serum albumin using gluteraldehyde (6).

V3:Ty-VLPs: Ty-VLPs containing the V3 loop sequence (as above) were constructed as previously described (7). The V3 loop sequence comprises 10.5% of the molecular weight of the Ty:V3 fusion protein.

rgp120: Insect cell-derived recombinant gp120, isolate IIIB, clone BH10 (American Bio-Technologies). There is one sequence difference (R → S) as shown underlined on the V3 sequence above.

Immunization: Immunogens were injected subcutaneously near the base of the tail in a volume of 0.1ml, either as an aluminium hydroxide precipitate or in phosphate buffered saline. 7 - 10 week old mice of the following haplotype were used:

C57BL/6 (H-2^b), BALB/c (H-2^d), C3H-Hej (H-2^k), DBA/2 (H-2^d)
Charles River;   also CBA (H-2^k), Olac.

Proliferation assay: Inguinal draining lymph nodes (DLN) were removed seven days following immunization and single cell suspensions prepared in RPMI-1640 medium (Life-Technologies) containing 5% foetal calf serum (Flow Labs.), glutamine at 2mM penicillin at 100 IU/ml and streptomycin at 100μg/ml (Life- Technologies). Cells were cultured at 500,000 cells per well in flat bottomed microtiter plates (Nunc). Cells were challenged, in triplicate with peptides and proteins for 5 days. Proliferative responses were determined by measuring the incorporation of tritiated thymidine (0.5μCi/well) (Amersham International) during the last 18 hours of culture. Cells were harvested onto glass fibre filters (934-AH, Whatman), using a Skatron cell harvester (LKB Wallac) and counted in a Beckman LS 5000CE beta counter. Results were expressed as mean disintegrations per minute (d.p.m.) and converted into stimulation indices as follows:

$$\text{STIMULATION INDEX (S.I.)} = \frac{\text{d.p.m. in stimulated cultures}}{\text{d.p.m. in control cultures}}$$

An S.I. of greater than 2 is considered positive by most authors. The coefficient of variation (SD/mean) in stimulated cultures was generally between 10-20% throughout the range of d.p.m.

RESULTS

Proliferative responses to V3 in four mouse strains: Four strains of mice (C3H-Hej, DBA/2, BALB/c and C57BL/6) were immunized with 50μg of V3:Ty-VLPs as an alum precipitate. DLN were then challenged in the proliferation assay with V3 peptide at concentrations ranging from 6.25-100µg/ml. Figure 1 shows that proliferative responses were seen in three of the four strains at all the peptide concentrations tested. Background responses were 554,1058,998 and 5890 d.p.m. respectively. Figure 3 also illustrates that responses can be obtained in vitro with V3 peptide concentrations as low as 0.1µg/ml. Normal mouse DLN cells gave no proliferative responses to V3 peptide over the dose range tested.

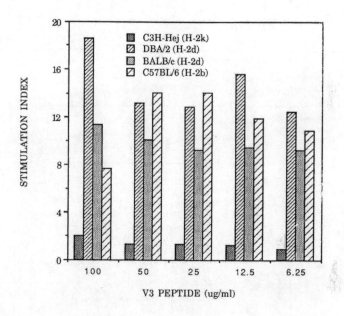

FIGURE 1. Proliferative responses to V3 peptide in strains of mice immunized with V3:Ty-VLPs.

Immunization dose response - V3:Ty-VLPs and the response to V3 peptide: V3:Ty-VLPs immunization doses of 0.3 to 175µg in alum were administered to BALB/c mice.Seven days following immunization DLN were removed and challenged in vitro with V3 peptide at 5µg/ml. An immunization dose dependent increase in proliferative responses was observed (Table 1), with significant responses seen at immunization doses as low as 7µg. Background responses were 4585, 832, 639, 100 and 114 d.p.m. for the groups 175 - 0.3µg respectively.

Proliferative responses to V3 peptide following immunization with either V3:Ty-VLPs, rgp120 or V3-alb: Three strains of mice (C57BL/6, DBA/2 and CBA) were immunized with either 15µg of V3:Ty-VLP 17.3µg of rgp120 or 11.7µg of V3-alb as an alum precipitate.

TABLE 1. Effect of V3:Ty-VLP immunization dose
on proliferative responses to V3 peptide.

IMMUNIZATION DOSE (μg) :	175.0	35.0	7.0	1.4	0.3
S.I. (5μg/ml V3) :	11.1	7.9	5.9	1.4	1.0

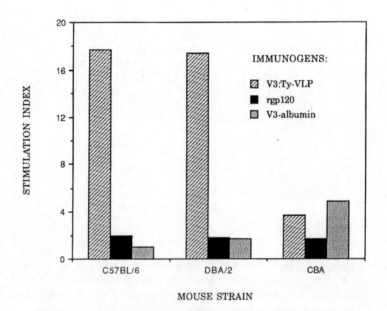

FIGURE 2. Proliferative responses to
V3 peptide following immunization with
equivalent doses of three immunogens.

These doses each represent 1.575μg of V3 peptide. DLN were removed at 7 days and challenged with V3 peptide at 0.01 - 10μg/ml. Figure 2 shows that immunization with V3:Ty-VLPs led to a greater proliferative response to the V3 peptide in C57BL/6 and DBA/2 mice at 10μg/ml than immunization with equivalent doses of rgp120 or V3-alb. Responses to V3 peptide in CBA mice were generally low, and were similar to those seen with C3H-Hej mice (of the same haplotype).

The background d.p.m. were as follows:

Immunogen	C57BL/6	DBA/2	CBA
V3:Ty-VLP	2338	4050	1100
rgp120	2658	13123	6530
V3-alb	670	410	198

The reason for the variation in background responses is not known.

Proliferative responses to rgp120 following immunization with V3:Ty-VLPs: Four strains of mice were immunized with 15μg of V3:Ty-VLPs as an alum precipitate. DLN were challenged at seven days with rgp120 at 1μg/ml. Proliferative responses were observed in C57BL/6, BALB/c and DBA/2 mice, with no significant response in the CBA group. The response to rgp120 is shown in Table 2.

Proliferative responses to V3 peptide following immunization with V3:Ty-VLPs, with and without adjuvant: C57BL/6 mice were immunized with 50μg of V3:Ty-VLPs as an alum precipitate or with no adjuvant. After 7 days DLN were challenged in vitro with V3 peptide. The responses are shown in Figure 3. Proliferative responses were seen in both groups although greater responses were seen when adjuvant was used.

TABLE 2. Proliferative responses (S.I.) to rgp120
following immunization with of 15μg V3:Ty-VLP.

STRAIN:-	RESPONSE TO rgp120 (μg/ml)		
	1.00	0.10	0.01
CBA ($H-2^k$)	2.6	2.0	1.3
C57BL/6 ($H-2^b$)	7.2	4.3	2.9
DBA ($H-2^d$)	3.7	3.5	N.T.
BALB/c ($H-2^d$)	7.7	4.8	N.T.

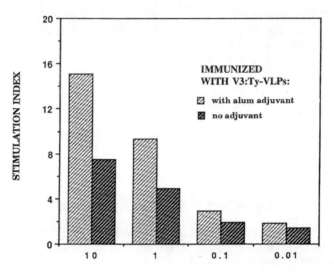

Figure 3. Effect of V3:Ty-VLP immunization
with and without adjuvant on proliferative
responses to V3 peptide.

DISCUSSION

We have demonstrated V3 sequence specific proliferative T-lymphocyte responses
following immunization of mice with V3:Ty-VLPs either as an aluminium hydroxide
precipitate or with no adjuvant. The responses were observed in two of three haplotypes
tested. These results are encouraging, firstly because aluminium hydroxide is the only
adjuvant currently licenced for human use, and, secondly, because a vaccine would need
to induce immunity in a broad range of human HLA haplotypes. The V3 reactive T cells
induced by V3:Ty-VLP immunization were also able to respond to recombinant gp120.
Although the conformation and glycosylation patterns of rgp120 are not identical to
HIV-gp120 (7), the T cell epitopes, based on amino acid sequences, will be unchanged
and we might expect such T cells to respond to HIV-gp120.

The presentation of the V3 sequence in a particulate form may confer an immunological
advantage. To investigate this we compared the V3-specific T cell responses obtained
following immunization with V3:Ty-VLPs to those seen after immunization with V3-alb
and rgp120. The conjugation of peptides to albumin is an accepted way of increasing

immunogenicity (6), and rgp120 is an alternative non-particulate protein presenting HIV V3 (3). When mice were immunized with doses of the three immunogens delivering the same dose of the V3 sequence, V3:Ty-VLPs gave consistently higher responses to V3 peptide. This would imply that this particulate system is more efficient in terms of the uptake, processing and presentation of T cell epitopes than the non-particulate. It is of interest to note that the only recombinant DNA sub-unit vaccines which have proved to be efficacious in humans are Hepatitis B surface antigen preparations, which also assemble into particulate structures (8).

The above results are based on a single 40mer sequence insert, but the Ty-VLP technology is adaptable and can accept a wide range of additional protein sequence without the disruption of particle formation (9). We are currently investigating the immunogenicity of Ty-VLPs carrying a number of additional T and B cell epitopes.

ACKNOWLEDGEMENTS

We wish to acknowledge the assistance of the following: Jo Griffiths, Kevin Jones, Nigel Burns, Nick Abbott, Jamie Biggins and Mark Cunningham for supplying purified Ty-VLPs, John Senior for preparing V3-alb and the Medical Research Council (ADP) for providing rgp120.

REFERENCES

(1) N. A. Hessol, A. R. Lifson, P. M. O'Malley, L. S. Doll, H. W. Jaffe and G. W. Rutherford, Prevalence, incidence and progression of human immunodeficiency virus in homosexual and bisexual men in hepatitis B vaccine trials. Am.J.Epidemiol. **30**:1167 (1989)

(2) D. D. Ho, M. G. Sarngadharan, M. S. Hirsch, R. T. Schooley, T. R. Rota, R. C. Kennedy, T. C. Chant and V. L. Sato, Human immunodeficiency virus neutralizing antibodies recognize several conserved domains on the envelope glycoproteins. Proc.Natl.Acad.Sci. U.S.A. **85**:1932 (1987)

(3) P. Berman, T. Gregory, L. Riddle, G. Nakamura, M. Champe, J. Porter, F. Wurm, R. Herschberg, E. Cobbs and J. Eichberg, Protection of chimpanzees from infection by HIV-1 after vaccination with recombinant glycoprotein gp120 but not gp160. Nature **345**:622 (1990)

(4) J. A. Berzosky, A. Bensussan, K. B. Cease, J. F. Bourge, R. Cheynier, Z. Lurhuma, J-J. Salaun, R. C. Gallo, G. M. Shearer and D. Zagury, Antigenic peptides recognized by T lymphocytes from AIDS viral envelope-immune humans. Nature **334**:706 (1988)

(5) H. Takahashi, J. Cohen, A. Hosmalin, K. B. Cease, R. Houghten, J. L. Cornette, C. Delisi, B. Moss, R. N. Germain and J. A. Berzofsky, An immunodominant

epitope of the HIV gp120 envelope glycoprotein recognized by class I MHC molecule restricted murine cytotoxic T lymphocytes. Proc.Natl.Acad.Sci. U.S.A. **85**:3105 (1988)

(6)   A. Johnson and R. Thorpe, "Immunochemistry in Practice", Blackwell Scientific Publications, London (1987)

(7)   J. R. Rusche, D. L. Lynn, M. Robert-Duroff, A. L. Langlois, H. K. Lyerly, H. Carson, K. Krohn, A. Ranki, R. C. Gallo, D. P. Bolognesi, S. D. Putney and T. J. Matthews, Humoral immune response to the human immunodeficiency virus envelope glycoprotein made in insect cells. Proc.Natl.Acad.Sci. U.S.A. **84**:6924 (1987)

(8)   M. J. Redmond and L. A. Babiuk, The application of biotechnology to animal vaccine development. Genetic Engineer and Biotechnologist **11**:14 (1991)

(9)   A. J. Kingsman and S. M. Kingsman, Ty: a retroelement moving forward. Cell **53**:333 (1988)

23

TEACHING IMMUNOLOGY: A MONTESSORI APPROACH

USING A COMPUTER MODEL OF THE IMMUNE SYSTEM

Franco Celada* and Philip E. Seiden**

*Department of Molecular Immunology, The Hospital for Joint Diseases, 301 East 17th Street, New York, NY 10003 USA. **IBM T.J. Watson Research Center, P.O. Box 218, Yorktown Heights, NY 10598 USA

INTRODUCTION

The revolutionary novelty of Maria Montessori's educational method was to furnish children with tools and materials and let them discover the hidden possibilities by *working them out* with their own hands. In 19th-century society, such an approach did not come "naturally" but had to be brought about by cultivation. In fact, the method has deep roots in western philosophy. In particular it stems from the very original thought of Giovanni Battista Vico (1668-1744), of the Neapolitan school (it is nice to mention him as a neighbor in connection with the course in Sardinia). Vico's strong point was that man's knowledge is limited. In fact any man could *really* know only what he has *made* himself. The connection between making and knowing is powerful, and in two hundred years has conquered the world, with a movement remarkably accelerated during the last decades. The recent blooming of personal computers and video games is not merely a fortuitous coincidence. To have a direct experience of the phenomenon it is sufficient to walk into a recently built museum of science, anywhere in the world, say, Hong Kong, where the impressive building was inaugurated in January 1992. There are no exhibits as such, but only curious machines waiting for the visitor to start them. The general atmosphere is set by a towering construction that occupies most of the four-story hollow space and is a three-dimensional automatic pinball machine with steel balls the size of pumpkins that kick off, turn, are deviated, hit flags and bells, fall into pits or rise to seemingly unnatural heights and roll around the building on their minimalist rails. The machines themselves are really "hands on": a rope to be grabbed which through a series of pulleys translates the little boy's force into the movement of an impressive stone; bimetallic plates that generate measurable electric currents -- proportional to heat and humidity -- when both palms are laid upon them. These are not real machines of everyday life, but they are not pure games, either. They belong to the intermediate world of pilot plants and Gedanken experiments. We wonder whether a similar approach would be feasible in the teaching of science to professional science students and in particular -- since this question arises in the midst of an advanced immunology course -- whether it could be applied to the teaching of immunology. Putting real experiments systematically in the student's hands would be the best but is utterly unrealistic in terms of numbers of pupils and equipment. Completely abstract games have obviously no meaning. We suggest an intermediate path which requires a workable model of the immune system that can be modified -- "made" -- at any moment, where each modification constitutes an experiment, and where each experiment leads to a little step of knowledge and triggers the next experiment.

In the conviction that this is a reasonable approach, we decided to illustrate our computer model of the immune system at the Porto Conte course, in the hope of obtaining indications as to whether the model can be a tool for teaching. In this article we shall explain the simulation and then test it by confronting it with some real immunological problems.

THE MODEL

The immune system is endowed with almost limitless complication. Obviously our model cannot contain all parts of the biological system, so we must make a selection. We have chosen to focus on the cellular and molecular interactions which trigger and regulate the humoral immune system. Three types of cells are considered, A, B, and T. The A cells represent professional antigen presenting cells. They are nonspecific scavengers of antigen, which they capture with low but uniform affinity (affinity being taken to mean the probability of binding antigen). In contrast, the B cells are endowed with specificity in the form of a binary receptor, i.e., a string of zeros and ones. As described below, these receptors will only bind to "matching" antigenic epitopes but will do so much more strongly than the A cells. The T cells are helper cells and also have a specific binary receptor. In addition to receptors we supply our B and A cells with major histocompatibility complex (MHC) molecules. These are also of binary form. Figure 1 shows a cartoon of the three types of cells.

Antigen epitopes are also taken as binary strings, and an antigen can contain as many epitopes as desired. In addition to the epitopes to which A and B cells will bind, our antigens must contain the peptides that will eventually bind to the MHC molecules as described below. In the actual biological case these are pieces of the antigen broken down by the antigen presenting cell. However, we do no biochemistry in our model, so that we must specifically denote the peptides as well as the epitopes constituting our antigens. Some typical antigens are shown in Fig. 2.

We also include antibodies which consist of a peptide and an epitope identical to the receptor of the B cell which produced it. It can bind to antigens by means of this epitope as the B cells does. However, in addition, all antibodies contain an Fc region and all A cells carry Fc receptors. This allows the antibody which binds antigen to be bound in turn to an A cell by means of its Fc region. This binding has an affinity midway between the B cell and A cell affinities for antigen.

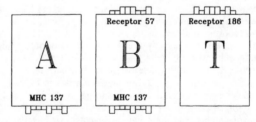

Fig. 1. The three types of cells included in the model. The binary receptors and MHC molecules are shown here for the specific case of 8-bit strings. The number given for them is the decimal value of the binary strings.

Antigen 198 228 25 218

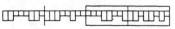

Fig. 2. A typical antigen showing its epitopes (open segments) and peptides (boxed segments) for the case of 8 bits per segment.

These are the components included in our model. The interactions take place as follows. Antigen introduced into the system may be bound by either the B cells or the A cells. The A cells may bind any antigen but with low affinity. The B cells may only bind matching antigens, i.e., antigens whose epitopes match the receptors of the B cells.

There are essentially two ways by which biological macromolecules interact specifically with each other (and are thus said to "recognize" each other). One is the attraction between identical stretches of amino acids, e.g., brought about by the juxtaposition of hydrophobic centers, and typically occurring between similar protomers making up a polymeric structure. The other way, which happens to be the basis of immune recognition at various levels, is complementary in terms of function and shape. That is, it involves the proximity of a positively charged group to a negatively charged one, magnetic interactions, and the so-called lock and key fitting of 3-dimensional shapes. We simulate the complementarity interaction by considering as a match a zero on the epitope aligned with a one on the receptor, or vice versa. We do not require perfect matches, i.e., all bits matching, but the affinity of the interaction decreases with the number of mismatching bits. Typically, for the 8-bit case reported here we allow 8- and 7-bit matches to bind, but the affinity for the 7-bit match is an order of magnitude less than the 8-bit match.

After the epitope is bound to an A or B cell it is processed and its peptides are presented on the MHC molecules of the cell. It is to this MHC/peptide complex that a T cell will attempt to bind. The T cell receptor sees both bare MHC and the peptide. We simulate this by letting the left half of the MHC represent the piece of bare MHC and the right half the peptide groove. We then take the antigenic peptides and try to bind them to the right half of the MHC. This involves one half of the peptide bits. When one half binds, the other half is presented to the T cell. A cartoon of this process is shown in Fig. 3.

When a T cell binds to a B cell in this manner, both cells are allowed to divide, establishing clones of cells having those particular receptors. In the case of a T cell binding to the MHC/peptide complex of an A cell, it is only the T cell which divides. Some of the newly produced B cells become plasma cells, producing antibodies with the same specificity as the B cell.

In addition to clonal growth, virgin B and T cells are continuously created from the "bone marrow". These are created with the full diversity possible. For a binary system the diversity is 2^n, where n is the number of bits. Therefore, for $n=8$ the diversity is 256. The T cells, however, are required to pass through a thymus before being allowed to participate in the process described above. In the thymus the T cells are subject to the same interactions as in the periphery, while being exposed to bare MHC and MHCs bearing self peptides. If the T cells bind, they are eliminated in proportion to their binding affinity. This is negative selection, and removes T cells which are dangerous because they exhibit a high affinity for self (and thus could cause autoimmune aggression). In addition we also have a positive selection step which chooses those T cells capable of binding at an intermediate level of affinity. This process removes those T cells that are not able to recognize self MHC at all and are therefore useless by Harald von Boehmer's definition (see his article in this volume).

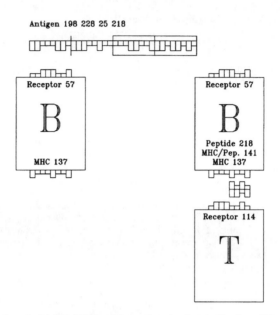

Fig. 3.   Steps in the recognition and processing of antigen by B cells and T cells. The decimal numbers are the values of the binary strings of the epitope, peptide, receptor, MHC and MHC/peptide complex.

These processes are implemented by using a computational technique called a cellular automaton (where the word cellular stands for "site" and not for cell in the biological sense).[1] A cellular automaton comprises a dynamic system consisting of discrete spatial dimensions and evolves in discrete time steps under local rules. The most familiar example is probably "Conway's Game of Life" which is played on a two-dimensional lattice (a checkerboard).[2,3] In our model we also represent the "body" by a two-dimensional lattice. All the entities described above populate the sites in this lattice. A schematic picture of four of these sites is shown in Fig. 4. The interactions are strictly local so that only entities in the same lattice site can interact. Each site is looked at separately and all possible

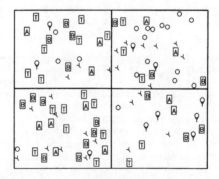

Fig. 4.   Four sites in the cellular automaton. The A, T, and B show three cell types. The circles are antigens, the Y-shaped figures are antibodies, and the circles attached to Ys are the antigen/antibody complexes.

interactions are considered for all entities in the site. The interactions that win out depend on their relative binding strengths. Each process is considered proportional to its binding strength or affinity and the winner is decided stochastically by means of a random number generator. This procedure is carried out for each site, then all the successful bindings and cell divisions take place simultaneously. New cells will be allowed to be born and old cells to die. Finally the entities are allowed to diffuse to neighboring sites. The totality of these processes is called a time step. The new distribution of entities is then considered again for the same processing on the next time step. The calculation is run for as many time steps as desired.

We have only outlined the basic ideas of our model here, as highly detailed descriptions have already been published.[4,5] Furthermore, this is only the basic model. We want to emphasize that the construction of this model is open-ended. New entities and interactions may be added quite easily as desired. For example, above we have only considered T helper cells containing class II MHC, but cytotoxic T cells with class I MHC could easily be added. Similarly, the interactions between entities can be added or modified as desired in order to test other aspects of the functioning of the immune system.

EXPERIMENTS

The most exciting feature of our model is that by running it repeatedly and varying one parameter at a time, for instance the input of antigen from outside, or the number of cells, or the interaction strength, etc., it is possible to perform real -- if simple -- experiments. We will call these experiments *in machina*: they are one step further removed from the biological reality as compared to *in vivo* and *in vitro* experiments. Experiments run *in machina* may serve two different purposes. The first is to verify the functioning of the model itself, recognize the similarity in behavior to a real immune system and serve as the basis to modify and improve it. The second is to go to situations which could never be attained in biology and "see what happens". What would happen if a population of 10,000 would evolve over 250 generations -- the selective trait being the immune efficiency of their major histocompatibility proteins? What would happen if the efficiency of thymus were lowered or the antigen dose increased up to 10,000 times the threshold? If the questions asked are not tossed up at random, but follow precise abductions (hypotheses), the results may be informative about the tenability of the latter even in real biology, since all the parameters incorporated in the model are biological. Obviously these experiments do not substitute for real facts but could orient conventional experimentation by indicating which are the more promising or less promising directions for the search.

In this article we want to demonstrate some *in machina* experiments of both types. To confirm that the model works as expected, i.e., as a bona fide immune system, we will present tests of immunization by antigen introduction and the effect of varying antigen doses over a wide range.

Characteristic of immunization is the building up of a specific response as a consequence of contact with a foreign substance (antigen) and the subsequent persistence of memory of this event, which results in a quantitatively and qualitatively modified response when the same antigen is introduced again. To test this in the model, we conducted an experiment where we inject a foreign 2-segment antigen (one epitope and one peptide) at time zero and then again at time step 40. Figure 5 shows the growth of the cellular and antibody population as a function of time.

The population of the B cell and T cell clones that respond to the foreign antigen grow to values much greater than the virgin populations. In addition the antibody population also grows to large values. At the second injection of antigen the antibody population is still large so most of the antigen disposal is effected by the existing antibodies. The effect on the B cell, T cell, and antibody populations is small, but the antigen population is reduced very rapidly.

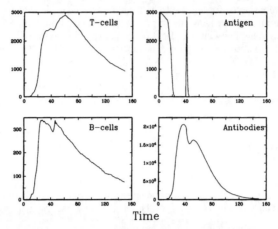

Fig. 5. Response of the model immune system to two injections of antigen. Antibody dominated secondary response.

One further point to be noted about Fig. 5 is that the T cell clonal growth starts before the B cell growth. This is more clearly shown in Fig. 6, where the cellular populations are shown for the first 25 time steps only. One can clearly see that the T cell population rises before the B cell population. This is due to the fact that the A cells are the first defense in a virgin system and that most of the antigen binding during the first 10-15 time steps is due to them. They will present the antigenic peptide to the specific T cells and the stimulated T cells will divide, creating a clone of T cells that can all bind to this particular antigenic peptide. At this point the specific B cells, which are much less prevalent in a virgin system than the non-specific A cells, will see their probability of meeting this specific T cell significantly increase. This *two-stage* process is important because it allows the replacement of a quadratic process with two linear processes. The direct clonal growth of B and T cells is proportional to a product of the populations of the appropriate B and T cells. Since these are exceedingly small in a virgin system one would have to wait for these chance meetings to start the clonal selection process. With a large population of A cells available the process is linear in small population variables.

As we noted above, the response of the system to the second injection of antigen was due to the large antibody population. In Fig. 7 the same experiment is shown but with the

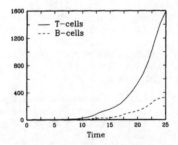

Fig. 6. Initial stage of clonal growth.

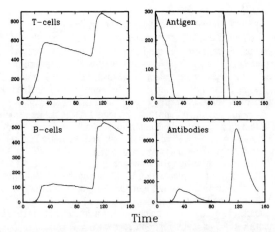

Fig. 7. Response of the model immune system to two injections of antigen. B cell
dominated secondary response.

second injection delayed until time step 100 when the antibody population has decayed close
to zero. In this case the response is due to the large population of selected memory B and
T cells and their interaction prompted by the antigen. As a result a second period of clonal
growth ensues and the population of cells increases even more. As soon as the antibody
population rises to a large value the antigen is rapidly removed.

Let us now figuratively take a magnifying glass and look at the quality of the primary
and secondary responses elicited by the introduction of antigen. The computer can easily
dissect the polyclonal response into its components, and show them in graphs like the one
of Fig. 8. As expected, nine clones of B cells take part in the response, one exhibiting an
8-bit match with the antigen epitope and the others with different 7-bit matches. The real
emerging feature is, however, the different behavior of the clones. As Fig. 8 shows, during
the primary response all 9 clones experience comparable proliferation and there is no
advantage for the high affinity clone. Instead, during the secondary response the high
affinity clone clearly takes the lead, thus causing the average affinity of the response to
increase.

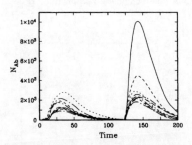

Fig. 8. Antibody response of the model system to two injections of antigen. The solid
line represents the high-affinity antibody (8-bit match). The other eight lines
show the individual low-affinity antibodies (7-bit match).

This result is very similar to the classical descriptions of primary and secondary responses *in vivo*. The fact that it is reproduced in the simulation where only a fraction of the factors and conditions operating in the living organism are incorporated is in keeping with the concept that nature likes redundancy, and usually utilizes several means to reach one end, but indicates that at least one efficient mechanism has been introduced in the model. We can analyze how the control of affinity differs between primary and secondary response. Obviously selection of affinity, on which the clonal selection scheme is operating, has a smaller impact the first time antigen appears. The model result can be explained by the fact that the cellular interactions eventually leading to antibody secretion follow two possible pathways: one is Ag -> A cell -> T cell -> B cell, and the other is Ag -> B cell -> T cell. What decides which of these pathways is followed in each "site" of the model is the relative prevalence of B cells specific for the antigen epitope, and the same factor is decisive for triggering a competition for antigen among B cells -- which will naturally favor the high-affinity binders. For the virgin system the number of B and T cells per site is small. This means that the probability of finding a high-affinity B cell and an appropriate T cell at the same site is very small. The situation is much better for low affinity B cells since there are many more of them (8 times as many for 8-bit segments). Therefore, although the affinity of a high-affinity B cell is much greater than a low-affinity cell they do not dominate the scene because they have difficulty in finding T cell help. At the second injection, however, the population of both the high-affinity B cells and appropriate T cells has grown to a large value so that there is little difficulty in finding help. Now the high-affinity B cells do dominate because of their greater affinity.

A second experiment for "testing the system" concerns the regulation of antibody production at high antigen doses. It is an old notion that by increasing the antigen dose injected, the antibody production first increases, then reaches a plateau, and finally decreases, producing a phenomenon that has been called high dose tolerance. Several hypotheses have been put forward to explain the phenomenon, e.g., saturation of antigen receptors, competition for antigens between memory cells and antibodies, enhanced network interactions, anergy of B cells, elimination of B and/or T cells. The use of the minimal model, where idiotypic interactions and B cell anergy are not operating, to mimic high dose tolerance, *de facto* isolates the first two of the mechanisms listed above and allows us to test whether receptor saturation and B cell-antibody competition alone can reproduce the phenomenon. The experiment was conducted by delivering antigen challenges varying between 10^1 and 10^5 molecules to a body that had been primed with a dose of 10^3, and recording the peak antibody titer of the subsequent secondary response. The data are expressed as the number of antibodies relative to the challenge dose, and are recorded separately for high-affinity and low-affinity antibodies. Fig. 9 shows that there is high dose regulation in the system, and also that the high-affinity response is more susceptible to this regulation, as the decline in relative antibody production begins at a 10-fold lower antigen

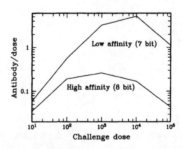

Fig. 9.   Response of the model system to the size of the challenge dose.

dose. These results mimic with remarkable accuracy observations made in adoptive responses *in vivo*.[6] They are particularly interesting because they indicate that the competition between antibodies and B cells is a central mechanism of regulation of the response.

We wish now to illustrate two experiments of the *exploring* type. The first one is concerned with self peptides and the delicate balance that the immune system must negotiate between inefficiency and autoimmunity, and essentially deals with problems similar to those discussed by N. A. Mitchison elsewhere in this volume.

We want to ask how large the self peptide repertoire may be relative to the external antigenic universe of peptides. We are obviously on rather soft ground, as so many factors are still unknown. For instance, the combinatorial possibilities of 10-unit peptides made with 20 amino acids are 10^{20}, an astronomical figure, but the effective peptidic repertoire is much smaller, as Lanzavecchia showed in this course, since only those peptides that can be effectively presented by MHC are immunogenic. A minimalist view concerning the self repertoire was heralded in Porto Conte by Irun Cohen, who would make it just large enough to give the immune system a sufficient if sketchy idea of the body: the homunculus. In our case the best we could do was to take the model's total number, which is 256, as the universe of peptides, and test what fraction of that number is compatible with a reasonably efficient capacity to respond to external antigens.

We proceed as follows. We choose a set of 100 random antigens from the full diversity possible. Then we set up a model system with one self peptide chosen randomly. We then expose the system to the set of random antigens and calculate the net response of the system to the antigens (responsivity). This is essentially the probability that the system will survive the attack by foreign invaders represented by the set of antigens. We then choose another random self peptide and repeat the calculation. We do it 100 times for 100 random choices of self peptide and average the results. This number is the average responsivity of a one self peptide system. The calculation is then repeated for 2 randomly chosen self peptides, then 3, 4, etc., up to 50. The results are shown in Fig. 10. The responsivity decreases with increasing size of self: at first slowly, then more rapidly, reaching a roughly exponential slope. This is mainly due to the shrinking of the T cell repertoire available for defense, imposed by the increasing severity of negative selection in the thymus. An approximate indication of how large a self pool is compatible with a reasonable capacity to respond can be gathered by extrapolating the exponential back to unity (the dashed curve in the figure). The intercept is at 15 peptides, or 6% of a repertoire of 256. It certainly is a surprisingly low figure. It will be interesting to compare this figure with biological estimates.

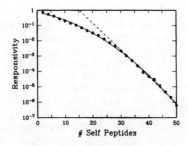

Fig. 10.   Responsivity of the model system as a function of the number of self peptides.

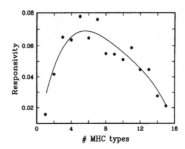

Fig. 11. Responsivity of the model system as a function of the number of different MHC types per individual.

The second *exploration* regards the MHC molecules. Their principal function is the presentation of antigen to the T cells. MHC class I interacts with $CD8^+$ T cells (mostly killer cells) and MHC class II interacts with $CD4^+$ T cells (mostly helper cells). In the present model only the latter relation is simulated.

In nature every individual carries a limited number of different MHC molecules and all of them are represented in each antigen presenting cell. We wondered whether the number of MHC types was an evolutionary product of the double function exercised by these molecules, one during thymic selection (in particular the negative selection) and the other during T cell stimulation in the mature individual. We therefore set up an experiment to determine the effect of varying the number of MHC types per individual on the efficiency of the immune defense. Responsivity was determined in the same way as for the experiment on the size of self, but in this case we vary the number of MHC molecules instead of the number of self peptides. The results (Fig. 11) show that responsivity is not optimal with a single MHC molecule. The availability of more types, up to 4-6, improves the responsivity, obviously by allowing a more efficient presentation of foreign peptides to the T cells. However, when the number of MHC increases over these values, the responsivity declines, the cause being the creation of more and more "holes" in the T cell repertoire by negative selection in the thymus. The general shape of the curve and its logic is pleasing. The almost perfect agreement of the peak responsivity with the observation *in vivo* must be taken purely as coincidental, given the momentous differences -- for instance -- of the size of the repertoires between the model and the biological reality.

CONCLUSION

Let's go back, after this short demonstration of what a working computer model is and can do, to the proposition boldly stated in the introduction to this chapter: that working and playing with computer models can become a useful didactic tool in immunology. In the introduction, we have underlined the hands-on, personal engagement aspect of modeling and working with computer models, and we think this is one of the strong points of the proposition. A second point is our conviction that the field of immunology has reached a size, a *corpus*, so large as to make it difficult for someone approaching it for the first time to understand which are the fundamental interactions that make the immune system behave like an immune system, and which are the consequences, the ramifications, the sidelines, the phenomenology. Every part of the *corpus* is important, no doubt, providing that the perspectives are perceived clearly. The recent history of the discipline shows many cases of lack of perspective, i.e., situations where a phenomenon or a body of observations attracts extraordinary attention and generates a wave of enthusiasm lasting for years before being reduced to the right proportions. The waves cause the production of excessive experimental work, excessive numbers of papers published, and -- a more serious consequence -- a number of researchers who may find themselves overspecialized or misfit when the wave recedes. Examples are easy to recall: the transplantation antigen function, the anti-tumor immunity, the suppressor-cell-centered regulation, the natural killer cell immunology. Note also that waves of fashion can be followed by waves of neglect (see, for instance, the case of suppressor cells), with equally disastrous consequences. One of us, FC, who has experience in teaching immunology, feels that he has been able to avoid overemphasis of some of these -- certainly important -- areas, and has earned Antonio Lanzavecchia's citation in his EMBO Gold Medal Review for having "set [to young collaborators] only two conditions: the territory should be in the center of immunology and it should deal with real immune responses, i.e., antigen-driven."[7] We believe that teaching with models and work by students on the model can have the effect of educating the sense of perspective, for the simple reasons that (a) the information on which any model works is limited and must contain the essential in order to function, and (b) when the student will try to rebuild his own version of the model he will be obliged to ask first the fundamental questions to avoid utter confusion, and this in turn will stimulate the search through books, bibliography, discussion and, yes, experimentation for the pivotal center of the system.

REFERENCES

1. P. Manneville, N. Boccara, G. Y. Vichniac, and R. Bidaux (eds). *Cellular Automata and Modeling of Complex Physical Systems*, Springer-Verlag (1989).
2. M. Gardner, *Sci. Amer.* 223:120 (1970).
3. M. Gardner, *Sci. Amer.* 224:112 (1971).
4. F. Celada and P. E. Seiden. A computer model of cellular interactions in the immune system. *Immun. Today* 13:56 (1992).
5. P. E. Seiden and F. Celada. A simulation of the humoral immune system. *J. Theor. Biol.* (in press, 1992).
6. F. Celada. The cellular basis of immunologic memory. *In:* Prog. Allergy, vol. 14, pp. 223-267, Karger, Basel (1971).
7. A. Lanzavecchia. *EMBO J.* 7:2549, 1988.

24

MODELLING T CELL MEMORY IN VIVO AND IN VITRO

Angela R. McLean

Zoology Department
Oxford University
South Parks Road
Oxford OX1 3PS UK

INTRODUCTION

The immune response is regulated by a complex set of cellular
interactions, cytokine signals and spatial structures. One response to such
complexity is to search for the key interactions and try to understand
those. In this paper mathematical modelling is used to investigate possible
regulatory interactions between T helper subsets during the course of an
immune response to a foreign antigen. The model is used to investigate
questions of immune dynamics such as: what is immune memory; why can it not
be re-created in vitro; why are some replicating antigens (e.g. some viruses)
never cleared from the body? These are all addressed as problems in
cellular population dynamics.

The main part of the paper is in two sections; the first describes the
model and its assumptions, and the second presents results in the form of
model simulations. Mathematical details are presented in an appendix.

MODEL

The model describes interactions between five populations; three sub-groups
of T helper cells, interleukin-2 and antigen. The T helper sub-groups are
naive, activated and memory cells. Figure 1 is a representation of the
interactions that are assumed to take place between these populations, and
the model's assumptions are listed in detail below. The model's equations
are presented in the appendix.

Assumptions

1)   Naive T helper cells migrate from the thymus at a constant rate, and
     have a constant, per cell death rate. Therefore in the absence of
     antigenic stimulus the population of naive cells is fixed in size.
2)   Upon antigenic stimulation naive cells become activated. Details of
     the uptake, processing and presentation of antigen are subsumed into
     the assumption that the rate at which naive cells become activated is
     proportional to the amount of antigen.
3)   Activated cells produce IL-2 and are driven by IL-2 to divide, giving
     two daughter memory cells.
4)   At low densities of activated cells the per cell rate at which
     activated cells divide is proportional to the amount of IL-2.
5)   At high densities of activated cells the per cell division rate
     saturates to a constant. This represents the assumption that some form
     of suppression stops activated T helper clones from growing too large.
6)   Activated cells have the same death rate (and therefore the same
     lifespan) as naive cells.
7)   Memory cells are more rapidly activated by antigen than naive cells,
     and are also subject to a background, low level activation rate. It is

T Lymphocytes: Structure, Function, Choices, Edited by F. Celada
and B. Pernis, Plenum Press, New York, 1992

this background activation (coupled with the suppression of assumption 5) that is responsible for the long term maintenance of memory clones. It could represent cross-reactive stimulation (Beverly 1990) or presentation of sequestered antigen (Gray & Skarvall 1988).

8) Memory cells have the same death rate (and therefore the same lifespan) as naive and activated cells.

9) IL-2 is only produced by activated cells and only acts upon activated cells.

10) IL-2 is lost through binding to activated cells (at a rate proportional to the number of activated cells) and through degradation.

11) In the absence of specific immunity antigen grows exponentially. This exponential growth represents the net effect of virus reproduction and non-specific immune clearance.

12) Specific clearance of antigen is at a rate proportional to the number of activated cells. This represents the assumption that B cells and CD8+ cytotoxic T cells act at a rate proportional to the amount of help they get from activated T helper cells.

This statement of the model's assumptions is complemented by the equations presented in the appendix. In those equations, the rate at which each of these interactions proceeds is governed by a model parameter. These parameters represent a link between the model and experimental work and their definitions are therefore described in Table 1.

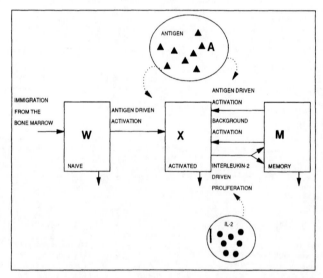

Fig.1   Model Structure. This model of the antigen driven activation and interleukin-2 driven proliferation of a clone of T helper cells counts five populations: naive, activated and memory T helper cells; interleukin-2 and antigen. Naive cells immigrate from the bone marrow, and can be activated by their specific antigen. Activated cells produce IL-2 and proliferate in response to IL-2 to produce two memory cells. Memory cells can be activated by antigen and are also subject to a background activation rate that is present even in the absence of antigen. Depending on the experimental system modelled, antigen may or may not be replicating, and it is removed at a rate proportional to the number of activated T helper cells.

Table 1. Model Parameters.

PARAM	BIOLOGICAL MEANING
Λ	immigration rate of naive T helper cells
μ	death rate of all T helper cells
$\alpha, \ \delta\alpha$	'per antigen' stimulation rate of naive and memory cells
ε	background stimulation rate of memory T helper cells
r	replicating antigen's growth rate
γ	per activated, specific T helper cell antigen removal rate
ρ/ξ	'per IL2' proliferation rate at saturating numbers of activated T helper cells
$1/\xi$	T helper cell concentration at half maximum T helper proliferation rate
ϕ	per activated T helper cell IL2 production rate
β	binding rate of IL2 to activated T helper cells
ψ	IL2 decay rate

RESULTS

The model is used to investigate T helper population dynamics in vivo and in vitro. First, the full model is studied to see if it can display behaviour akin to in vivo immune memory. Then results from a reduced version of the model (one where some of the parameter values are set to zero) are compared to some in vitro data on T cell population dynamics. In the absence of numerical estimates of the parameters in table 1, the model's behaviour is studied by characterising all possible types of behaviour. This is made easier by the mathematical technique (described in the appendix) of expressing the model in non-dimensional terms. A consequence of using dimensionless parameters is that 'time' in analyses and numerical simulations is expressed in terms of a model parameter (e.g. T cell lifespan) so cannot easily be related to ordinary units of time (e.g. weeks, days). For this reason the x axes in figure 2 have no units of time. The x axes in figure 3 have been labelled consistently across all simulations using units of days to allow easy comparison with the experimental data.

In vivo results

Figure 2 summarises the possible responses of this model clone of T helper cells to a replicating antigen. The parameter values in all three parts of figure 2 are the same except for the growth rate of the replicating antigen (parameter r in table 1 and in the appendix). Depending on the value of the growth rate there are three possible responses.

Figure 2a shows how an antigen with a low growth rate is cleared leaving immune memory. A small amount of replicating antigen is introduced at time zero, when all cells present are naive cells. In the absence of a specific immune response the antigen initially grows exponentially. The presence of the antigen drives naive cells to become activated (hence the transient fall in the number of naive cells) and the activated cells go through several rounds of division. Each time an activated cell divides, two memory cells are formed, and these are re-activated immediately by antigen. Thus the presence of antigen causes a rise in the number of activated and memory cells. Activated cells orchestrate the specific immune response to the antigen so that the amount of antigen falls and is eventually completely cleared. However, in the process of expanding in

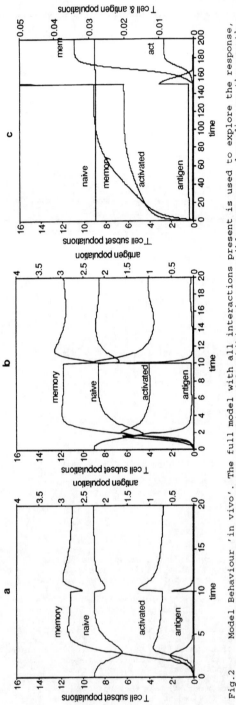

Fig.2 Model Behaviour 'in vivo'. The full model with all interactions present is used to explore the response, in vivo, to a replicating antigen. This model exhibits three possible responses depending on the growth rate of the antigen. (a) Replicating antigens with intermediate net growth rates are cleared leaving immune memory that will be maintained indefinitely. (b) Antigens with high growth rates cannot be cleared and instead establish low level, persistent infections that are never cleared. (c) There is a third group of very slowly replicating antigens that are also able to establish persistent infections if their infecting dose is very low. However these slowly replicating persistent antigens can be cleared by administering a comparatively large amount of antigen. This clearance is followed by long-lived immune memory. In (a) and (b) the left axis applies to the size of the T cell subset populations (in arbitrary units) and the right axis applies to the size of the antigen population (in arbitrary units). In (c) memory and activated T cell subset populations are of sizes as indicated on the right axis for time 0-150 units and as indicated on the left axis for time 150-200 units. (The simulations shown in (a), (b) and (c) were performed using composite parameters as described in the appendix. The parameter values used were the same for each figure for all parameters except antigen growth rate c. q=9, a=100, b=5, d=2, e=1.5, w=2, v=1, (a) c=4, (b) c=2, (c) c=0.02. Initial conditions were w=9, X=0, M=0, I=0, with A=0.01 for (a) and (b) and A=0.001 for (c).

response to the antigen, the activated and memory cells exceed the threshold above which they form a self-sustaining population. In this state background stimulation of memory cells (assumption 7) balances with suppression of the proliferation of activated cells (assumption 5) to generate a stable population of activated and memory T helper cells. This population is able to maintain immunological memory indefinitely. This memory is illustrated by the introduction of a large challenging dose of antigen at time 10 units, which is immediately cleared.

The size of the activated and memory population maintained in the absence of the replicating antigen depends only on internal interactions amongst elements of the immune system. This is why replicating antigens that grow too fast cannot be cleared and instead establish persistent infections. The response to such an antigen is illustrated in figure 2b. The course of events is the same as in fig 2a except that, instead of being cleared, the antigen population is reduced to low levels. This low level persisting antigen cannot be cleared even if the immune response is boosted with a large dose of antigen – as illustrated at time 10 units in figure 2b.

Figure 2a shows that the assumptions on which this model is built are sufficient to generate long term immune memory in the absence of persisting antigen. The 'cost' of suppression at high densities of activated cells is illustrated in figure 2b - there will be some fast replicating antigens that cannot be cleared. These two possibilities are intuitively appealing, but not very surprising. The unexpected prediction of this model is the existence of a third set of antigens with very low growth rates, which, if introduced at a low dose, can also establish persistent infections. One is illustrated in figure 2c. These antigens grow so slowly that they only stimulate their specific T cell clone slightly. The activated subset of the clone grows just enough to halt the further growth of the antigen, but never exceeds the threshold above which the activated and memory cells form a self sustaining population. However, a boost to the immune response with a 'large' dose of antigen (the large dose of antigen in figure 2c is the same size as the initial dose of antigen in figures 2a and b) is enough to push the activated and memory cell subsets over the threshold. In consequence the antigen is cleared leaving immune memory in the same manner as in figure 2a. This third possibility shows how the study of a mathematical representation of a set of biological assumptions can yield unexpected results.

In vitro results

The behaviour of this model described so far is very similar to that of a previously published model (McLean & Kirkwood 1990). However this earlier model makes no distinction between the situation in vivo and in vitro, so cannot be compared with in vitro data. To compare the current model with in vitro data, three of the parameters are set to zero - equivalent to switching off three of the interactions. The most important of the three is the background stimulation rate for memory cells. This is set to zero representing the assumption that, in vitro, either there is no cross-reactive stimulation of memory cells or there is no presentation of sequestered antigen. The other two differences are that there is no influx of naive cells and the antigen does not grow. Under these assumptions the model no longer displays immune memory and a single exposure to antigen leads to a transient activation and proliferation. Over the course of this response all cells convert from naive to memory phenotype. Figure 3 compares in vitro experimental results with simulations.

This model reconciles the existence in vivo of long term immunological memory with the observation that in vitro cultures of T helper cells need to be re-exposed to antigen every few weeks if they are to maintain proliferation. This reconciliation is achieved through the background stimulation of memory cells - assumed present in vivo, but absent in vitro.

CONCLUSIONS

A set of biological assumptions about interactions between three subsets of T helper cells, IL-2 and an antigen have been written as a mathematical model of the clonal dynamics of T helper cells. This model describes the antigen driven activation and IL-2 driven proliferation of a clone of T helper cells. Using two sets of parameter values appropriate to

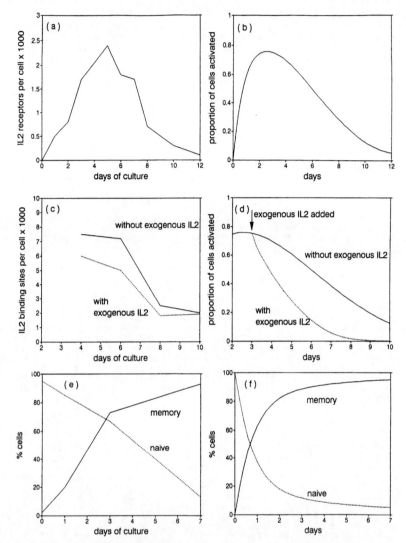

Fig.3　　　　Model behaviour 'in vitro' compared with in vitro experiments.
To compare the model's behaviour with in vitro experiments, a
number of the possible interactions in the model are switched
off by setting some parameters to zero. Most importantly there
is no longer any background stimulation of memory cells (e=0);
also there is no immigration of naive cells from the bone marrow
(q=0) and the antigen is assumed not to replicate (c=0). Time
series data on IL-2 binding sites per cell (a and c) are
compared with simulations (b and d) by plotting the proportion
of cells that are activated: in this model only activated cells
respond to IL-2. Both data and simulations show a transient rise
in cells able to respond to IL-2 (a and b) which can be
foreshortened by the addition of exogenous IL-2 (c and d). Over
the course of this transitory response, cells convert from being
mostly naive to being mostly memory (f) as has been demonstrated
in vitro (e). In vitro data are from (a) (Smith 1988) (c)
(Cantrell & Smith 1985) (e) (Akbar et al 1988). The following
parameter values were used in the simulations.
q=0,a=100,b=5,c=0,d=10,e=0,w=2,v=1.

the situation in vivo and in vitro the model has been used to explain why
immune memory can be generated in vivo but not in vitro. Further study of
the in vivo situation has revealed three possible types of model behaviour
akin to immune memory, persistent infection, and slow growing persistent
infection cleared after a boosting dose of antigen. The model's behaviour
with in vitro parameters has been compared to in vitro experimental results
and shown to be very similar. In material presented elsewhere (McLean 1992)
the model has been used to consider the possible impact of the reversion of
memory cells to naive phenotype, and as a basis for predicting the outcome
of some possible transfer experiments. What then remains to be done?

This model predicts that T helper cell memory is not just long-lived,
but indefinite. Experimental studies of the longevity of antiviral T cell
memory (Zinkernagel, 1990) suggest that protection may, in fact, be rather
short lived. Models in which T helper cell memory is still long lived but
slowly decays over time are currently under investigation. Studies using an
adaptation of this model to study anergy are also currently under way.

The reason for working with mathematical models like the one presented
here is that they can give precise and uncompromising representations of a
set of biological assumptions. Once the assumptions have been made into a
mathematical model _every_ consequence of those assumptions can be
investigated. So, for example, in looking for a set of assumptions that
would allow for the generation of long lived T cell memory, the study
presented here has shown that those same assumptions predict the existence
of antigens that grow too slowly to be cleared. This prediction could guide
further studies of viral persistence in vivo. This unexpected result
illustrates the benefits of precise exploration of immunological ideas
through mathematical modelling.

REFERENCES

Akbar A.N., Terry L., Timms A., Beverley P.C.L., & Janossy G.(1988) Loss of
    CD45R and gain of UCHL1 reactivity is a feature of primed T cells. J.
    Immunology **140** 2171-2178
Beverly P.C. (1990) Is T-cell memory maintained by crossreactive
    stimulation? Immunology Today **11** 203-205
Gray D. & Skarvall H. (1988) B cell memory is short-lived in the absence of
    antigen. Nature **336** 70-73
McLean A.R. (1992) Modelling T Cell Memory (submitted to International
    Immunology)
McLean A.R. & Kirkwood T.B.L. (1990) A Model of Human Immunodeficiency
    Virus Infection in T helper Cell Clones. J. theor. Biol. **147** 177-203
Smith K.A. (1988) Interleukin-2: Inception, Impact, and Implications.
    Science **240** 1169-1176
Smith K.A. & Cantrell D.A. (1985) Interleukin-2 regulates its own receptors
    Proc Nat Acad Sci USA **82** 864-868
Zinkernagel R.A. (1990) Antiviral T cell memory? Curr Top Micron Immunol.
    159 65-77

APPENDIX

The model counts five populations: resting Th cells (W), activated Th cells
(X) memory Th cells (M) interleukin 2 (I) and antigen (A).

Naive Th cells (W) migrate from the thymus at constant rate Λ and (if not
activated) have half-life $1/\mu$. In the presence of their specific antigen
(A) they become activated at a per-capita rate proportional to the amount
of antigen present αA. These assumptions give rise to the following
equation for resting Th cell dynamics:

$$\dot{W} = \Lambda - \alpha AW - \mu W$$

Activated Th cells (X) can be stimulated by interleukin 2 (I) to divide to
become 2 memory cells (M) this rate of division saturates at high levels of
activated Th cells. Activated Th cells have the same half-life $1/\mu$ as all
other Th cells. Memory cells can return to the activated state for two
reasons. First, presentation of the specific antigen (A) by APCs to memory
cells leads to their activation at a rate that is faster by factor δ than
the activation of naive cells. Second, a background activation rate ε is
always present. This could represent sequestrated antigen or cross-reactive

stimulation. These assumptions give rise to the following equation for the dynamics of activated Th cells:

$$\dot{X} \;=\; \alpha AW - \frac{\rho IX}{1+\xi X} + (\delta\alpha A+\varepsilon)M - \mu X$$

Activated Th cells divide to become two memory cells (M). These have the same half-life $1/\mu$ as other Th cell sub-sets. As stated above they are activated at per-capita rate $\delta\alpha A+\varepsilon$ representing more rapid activation by specific antigen plus some activation that is present even in the absence of specific antigen. The equation for the dynamics of memory Th cells is therefore:

$$\dot{M} \;=\; \frac{2\rho IX}{1+\xi X} - (\delta\alpha A+\varepsilon)M - \mu M$$

Interleukin 2 is produced by activated Th cells, absorbed by activated Th cells and has half life $1/\psi$. The equation for the dynamics of interleukin 2 is therefore:

$$\dot{I} \;=\; \phi X - \beta IX - \psi I$$

The equation used to describe the dynamics of specific antigen (A) depends on the type of antigen that is modelled. For a replicating antigen growing at rate r in the absence of specific immunity and removed at a rate that is proportional to the number of activated Th cells specific to it, the relevant equation is:

$$\dot{A} \;=\; rA - \gamma AX$$

The model has been reduced to dimensionless form in order to perform the numerical simulations - this reduces the number of parameters and the units used in analysis become unimportant (Murray, 1989). The reduction is performed by choosing new scaling variables for W, X, M, I, A and time t. Defining $W=W'W_0, X=X'X_0, M=M'M_0, I=I'I_0, A=A'A_0, t=t't_0$ and choosing:

$$W_0=\frac{\mu}{\gamma} \quad X_0=\frac{\mu}{\gamma} \quad M_0=\frac{\mu}{\gamma} \quad I_0=\frac{\mu}{\rho} \quad A_0=\frac{\mu}{\alpha} \quad t_0=\frac{1}{\mu}$$

the equations for the primed variables are in dimensionless form:

$$\dot{W'} \;=\; q - W' - A'W'$$
$$\dot{X'} \;=\; A'W' - \frac{I'X'}{1+vX'} + dA'M' + eM' - X'$$
$$\dot{M'} \;=\; \frac{2I'X'}{1+vX'} - dA'M' - eM' - M'$$
$$\dot{I'} \;=\; w(aX' - bI'X' - I')$$
$$\dot{A'} \;=\; A'(c - X')$$

where the composite parameters are given by:

$$q=\frac{\gamma\Lambda}{\mu^2}, \quad a=\frac{\rho\phi}{\psi}, \quad b=\frac{\beta\mu}{\gamma\psi} \quad c=\frac{r}{\mu} \quad d=\delta \quad e=\frac{\varepsilon}{\mu} \quad w=\frac{\psi}{\mu} \quad v=\frac{\xi\mu}{\gamma}$$

Values for these composite parameters used in the numerical simulations are given in the figure legends.

25

OLIGOCLONAL T CELLS EXPRESSING THE TCR V-ß 6 GENE PRODUCT FOLLOWING

IL-2 CULTURE OF OVARIAN CARCINOMA DERIVED LYMPHOCYTES

Eva Halapi[1], Mahmood Jeddi-Tehrani[1], Johan Grunewald[1], Roland Andersson[1], Christina Hising[2], Giuseppe Masucci[2], Håkan Mellstedt[2] and Rolf Kiessling[1]

[1]Department of Immunology, Karolinska Institutet and [2]Department of Oncology (Radiumhemmet), Karolinska Hospital, Stockholm, Sweden

INTRODUCTION.

T-cell populations obtained from tumor tissues (Tumor infiltrating lymphocytes, TIL) have been proposed to be enriched in lymphocytes specific for the autologous tumor. Experimental models have shown that TIL expanded in recombinant interleukin 2 (IL-2) are 50-100 times more effective in their therapeutic potency than IL-2-expanded LAK-cells derived from peripheral blood[1]. In clinical trials, therapies based on TIL combined with IL-2 have however so far yielded disappointing results, with response rates not better than those reported for LAK cells in combination with IL-2[2,3,4]. Also, it has generally been difficult to demonstrate a specific effector function of TIL against the autologous tumor, and IL-2 activated TIL from most types of tumors showed non-specific cytotoxic activity similar to that of LAK cells[5,6,7]. Moreover, there is a considerable practical problem in obtaining a sufficient number of TIL to be used for therapy.

Due to these considerations, a better understanding of the composition of TIL, with regard to subsets of T cells and NK-cells, would seem necessary. One important question to analyse, is whether TIL, or T-cell lines derived from these, contain dominant sets of T-cell clones. In one recent report, T cells from uveal melanoma were found to mainly express the V-α 7 gene product[8], although an analysis of the degree of clonality in this population was not performed. One could assume that T-cell clones which have been pre-activated and expanded by specific tumor-antigens should occur at a higher frequency, and express IL-2 receptors. This would lead to a more rapid expansion in the presence of IL-2 in vivo or in vitro. In the present report, we have investigated whether IL-2 expanded short-term lines of T cells derived from patients with ovarian carcinomas show restricted clonality, as measured by the use of a panel of anti-T-cell receptor (TCR) monoclonal antibodies (mAbs) and by studying the J-ß gene segment usage among T cells within certain V-ß gene families.

MATERIALS AND METHODS

Lymphocyte separation and culture: Tumor tissue was obtained from patients undergoing surgery. The aseptically removed tumor was cut in smaller pieces and further digested using an enzyme mixture consisting of 200U/ml of Collagenase IV, DNAse 270U/ml and hyaluronidase 35U/ml (Sigma, MO, USA) in a total volume of 20 ml for 3 hrs at 37°C. After washing in phosphate buffered saline (PBS) the cells were separated on Ficoll density centrifugation (Pharmacia, Sweden). Lymphocytes and tumor cells were recovered from the interface and washed 3 times in PBS. Ascites fluid from patients with ovarian carcinoma was centrifuged and the cell pellet was resuspended in PBS and fractionated on Ficoll. The interface was collected, and after 3 washes in PBS, the cells were further separated by centrifugation over a gradient consisting of 7 parts Ficoll and 3 parts PBS. Tumor cells were recovered from the interface and lymphocytes from the pellet. Lymphocytes (2×10^6/ml) were put in culture in RPMI 1640 (GIBCO, Paisley, U.K) supplemented with 100U penicillin, 100 μg/ml streptomycin ,4% human AB+ serum and 5 U/ml rIL-2 (kindly provided by Drs. J. Linna and J. Whitney, DuPont-Neymors).

T Lymphocytes: Structure, Function, Choices, Edited by F. Celada
and B. Pernis, Plenum Press, New York, 1992

Flow cytometry analysis: Phenotypic determination was performed by direct staining with phycoerythin-(PE)-conjugated mAbs to CD4 (Leu-3) or CD8 (Leu-2), purchased from Becton Dickinson, CA., USA,. and FITC-conjugated anti V-ß6 mAb (T Cell Sciences, MA., USA) All mAbs were used according to the manufacturers'recommendation. $5x10^3$ cells were analysed on a FACScan flowcytometer (Becton Dickinson). For detection of TCR V (variable) gene usage by TIL or peripheral blood lymphocytes (PBL), an indirect staining was performed with mAbs reactive to V-ß5.1,-ß5.2+5.3,-ß5.3, -ß6, -ß8, -ß12, and -α2.3 (T Cell Scienc) followed by FITC-conjugated rabbit-anti mouse Ig F(ab)2 (Dakopatts, Denmark)). Antibody reactive with V-α12.1[9] was a kind gift from Drs. M. Brenner and H. DerSimonian.
Preparation of RNA and cDNA synthesis: Total RNA was prepared from TIL and PBL by the guanidinum thiocyanate (GITC) method[10] . For first strand cDNA synthesis, 5 µg of denatured RNA (5 min. 95°C) was added to a reaction mixture containing dNTP (Pharmacia, Sweden), nucleotide hexamer (Pharmacia) reverse transcription buffer, dithiothreitol, Superscript Reverse transcriptase (BRL, MD., USA) and RNase inhibitor (Promega, WI., USA) according to recommendations of supplier and incubated for 45 min at 40°C and final denaturation of enzyme 5 min at 95°C.
Amplification of cDNA by PCR: 1 µl of cDNA was mixed with PCR mixture containing dNTP (200 µM each), 0.5 µM specific oligonucleotide[11], 0.5 U Taq Polymerase and 10x reaction buffer (Perkin Elmer Cetus, CT., USA) in a final volume of 20µl. The PCR profile used was as follows; denaturation 95°C 60 s, annealing 55°C 60 s and extension 72°C 60 s, for 35 cycles followed by final extension 72°C for 9 min. on a DNA Thermal Cycler (Perkin Elmer Cetus). Amplified products were applied to Hybond-N nylon filters (Amersham, U.K.), using a Bio-Dot SF microfiltration unit (Bio-Rad, CA., USA).
Jß gene segment usage: A detailed description of the method used for analysis of Jß gene usage has been reported elsewhere[12] and can be summarized as follows: C-ß and J-ß specific oligonucleotide probes were purchased (Scandinavian Gene Synthesis, Sweden) and end-labelled with r^{32}P-ATP (Amersham) for hybridization of amplified PCR material. After pre-hybridization of filters at 42°C in 2xSSPE, 5xDenhardts and 0.5% SDS, hybridizations were performed over night at 42°C with $1x10^6$ cpm/ml of the appropriate C-ß or J-ß specific probe. After being washed for 10 minutes at 42°C (in 0.2 x SSPE and 0.5% SDS) filters were exposed to Hyperfilm-MP (Amersham) overnight at -70°C. Gelscanning technique (2400 GelScan XL, Pharmacia-LKB) was used for quantification of exposed films. A standard curve was obtained by blotting known amounts of ^{32}p-labelled oligonucleotides, and used for a relative quantification of the blotted material.

RESULTS

Analysis of TCR V-gene usage in T-cell lines derived from ovarian carcinomas using mAb specific for V-gene products

A panel of mAbs specific for the V-α (a total of 2) or V-ß (a total of 6) families of the TCR-genes was used to analyse short-term T-cell lines derived from 9 ovarian cell carcinomas (Table 1). The flowcytometer analysis showed that 2 (M.M. and T.G.) of the 7 ascites derived T-cell lines, and 1 (E.S.) of the 2 lines derived from solid tumors expresses abnormally high frequencies (a value two times higher than the max. value obtained from the normal population[12]) of V-ß6 gene product compared to values obtained from healthy blood donors.The T-cell line of patient E.S. also had an increase in the number of cells positive for anti-V-ß5b (reacting with V-ß5.3) as well as for anti-V-ß12. None of the other 6 examined T-cell lines from tumor-derived lymphocytes showed enhanced levels of positive cells for other V-gene specific mAbs. A T-cell line was also established from the peripheral blood taken during the operation of patient E.S. The T cells in this cell line only contained a very low number of V-ß6+ T cells (Fig.1), in contrast to the high frequency of cells expressing this TCR gene in the TIL-derived cell line (38%). Neither did any of the other V-gene family specific mAbs show an increased number of positive cells in this PBL-derived cell line.

Analysis of the V-ß6+ subset in TIL cells

The V-ß6 subsets in the T-cell lines from patients E.S. and M.M. were analysed with double staining for co-expression of the CD4 and CD8 markers (Fig.2). As can be seen in figure 2, the majority (25.5% of the CD4+ cells in patient E.S. and 15.4% of the CD4+ in patient M.M.) of the V-ß6+ T cells in these lines were confined to the CD4 subset, and only a minor part (< 5%) of the V-ß6+ T cells co-expressed CD8. Freshly isolated tumor derived-lymphocytes from patient M.M. contained 5.6 % of V-ß6+ T cells (data not shown), which after 10 days of in vitro culture had increased to 15.4% of the total CD3 population. This increase in V-ß6+ cells only occurred in the CD4 subset, whereas the percentage of CD8+Vß6+ T cells decreased.

236

Table 1

Analysis of TCR V-gene cell surface expression on T-cell lines from ovarial cancer

Patient	Origin of tumour	CD3	αβ*	Vβ5.1	Vβ5a	Vβ5b	Vβ6	Vβ8	Vβ12	Vα2	Vα12
BL	Ascites	37	62	2.7	2.7	5.4	8.1	2.7	#	0.8	5.4
UL	Ascites	98	99	-	7.7	4.8	2.4	-	-	4.8	3.5
VG	Ascites	45.7	100	2.6	3.7	6.3	3.3	5.7	2.2	7.6	7.4
GE	Ascites	43.0	100	1.6	0.9	0.2	0.5	1.6	0.9	1.9	1.4
HN	Ascites	·72	-	0.4	1.4	0.0	5.8	3.5	0.4	1.4	0.6
MM	Ascites	92.9	94.5	8.5	5.2	1.5	**20.3**	2.9	0.9	2.7	9.3
TG	Ascites	80.0	100	7.2	1.2	0.2	**14.8**	1.9	2.9	9.4	2.4
ES	Solid	74	-	0.9	4.0	**17.2**	**38.4**	1.1	11.6	-	-
DS	Solid	18	-	2.8	3.9	0.0	4.4	16.1	0.6	3.3	0.6

(1.9-5.0) (2.1-8.7) (0.5-6.8) (0.3-7.8) (2.9-10.0) (1.2-2.5) (2.2-5.8 (1.3-5.5)

* Expressed as % of CD3+

Not tested

Values in brackets represents range in PBMC from normal donors (n = 24).[13]

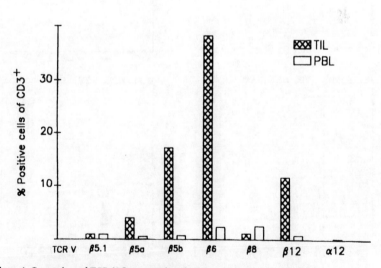

Figure 1. Comparison of TCR V-ß gene products in TIL and PBL from patient E.S.
Peripheral or tumor-derived lymphocytes were cultured for 20 days in IL-2 and analyzed by FACS with mAbs reactive with indicated V-gene products.

The question whether the V-ß6⁺ TIL cells were derived from a restricted number of T-cell clones was analysed using a recently developed method in which the the pattern of usage of J-ß gene segments was studied with a PCR based technique[10]. Figure 3 shows the "relative ranking" of the J-ß genes used among the V-ß6⁺ T cells from the TIL-line of patient E.S. The result is expressed as the contribution (%) of each Jß gene segment to the entire J-ß repertoire. As can be seen, the V-ß6⁺ TILs from this patient mainly used the J-ß2.7 gene segment. In contrast, the V-ß6⁺ T cells from a PBL line of patient E.S. cultured in parallel with the TIL line, revealed a different pattern with no J-ß gene being dominantly used.

When the V-ß6⁺ cells derived from the ascitis of patient M.M. were analysed by the same method, there was also some restriction with regard to J-ß gene usage, although this was not as pronounced as with the TIL line from patient E.S.(Figure 3 b). Four of the 13 J-ß gene segments were markedly overrepresented after 10 days of in vitro culture, with the J-ß2.1 and J-ß2.3 segments being the dominant ones. The non-cultured V-ß6 expressing cells were less restricted in their J-ß usage, although also in this population the J-ß2.1 segment was the most commonly used one among the V-ß6⁺ cells. This demonstrates that the IL-2 cultured TIL-cells from this ovarian carcinoma patient have a high proportion of V-ß6 positive cells which, in contrast to the PBL derived cells, appear to be derived from a very restricted set of T-cell clones.

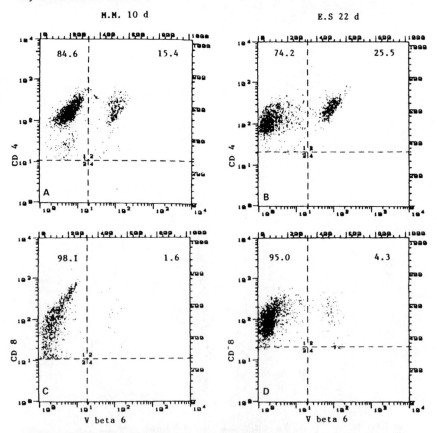

Figure 2. Analysis of V-ß6 expression in CD4 and CD8 subpopulations.
Double staining of M.M. day 10 in vitro culture (A and B) and E.S. 22 days in vitro culture (C and D) with FITC conjugated anti-V-ß6 and PE conjugated anti-CD4 (A and C) or anti-CD8 (B and D). By gating forward scatter/ fluorescence 2 only CD 4 or CD8 positive cells were analysed. Percent positive cells is given for each population.

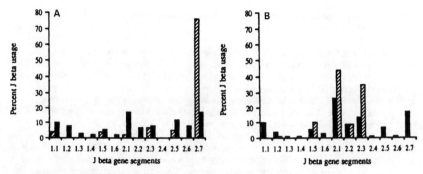

Figure 3. Percent J-ß gene segment usage among TIL and PBL.
A: Comparison of PBL (closed bars) and TILs (hatched bars) from patient E.S.
B: Comparison of TILs from patient M.M. on day 0 (closed bars) and grown for 10 days in low dose of
IL-2 (hatched bars).

DISCUSSION

This study demonstrates that a high proportion of IL-2 expanded T-cell lines derived from ascites or
solid ovarian carcinomas contained T cells with restricted TCR V-ß gene usage. It is of particular interest
that all 3 patients who showed enhanced levels of TCR V-ß gene expression, as measured by the panel of
mAbs used, had high proportions of V-ß6 + T cells. The majority of the V-ß6 + T cells were found in the
CD4 subset. Notably, the VB6 gene segment is also under normal conditions preferentially expressed by
CD4 + peripheral blood T cells as compared to the CD8 + T cells[13,14]. Freshly isolated tumor-derived V-
ß6 + T cells also expressed the CD45RO and CD29 markers (data not shown), while expression of
CD45RA was low, as analysed by flowcytometry. CD45R0 has been claimed to be associated with
activation and memory functions of T cells[15]. Although no clear correlation between cell surface markers
and function exists, since it has been shown that CD45 isoform expression is not unidirectional[16], it is
likely that expression of CD45R0 on TIL could reflect activation of these cells in vivo.

The yield of lymphocytes from solid tumors or ascites derived lymphocytes used in the present study
was for most patients too low to be analysed for V-gene usage by mAb, and only after 2-3 weeks of in
vitro IL-2 expansion was this possible. For one of the three patients (M.M.), we were able to obtain
sufficient numbers of tumor-derived lymphocytes for studies of freshly isolated T cells. Since the % of V-
ß6 + cells among these T lymphocytes only was one third of what was observed after 10 days of in vitro
culture, we conclude that the high proportion of V-ß6 + cells in the tumor-derived T-cell line reflects rapid
in vitro expansion rather than pre-existing high levels of V-ß6 + T cells.

In vitro culture of peripheral blood derived T cells in IL-2 did not select for outgrowth of V-ß6 + T cells
in this study. While a high proportion of V-ß6 + T cells were present in the TIL culture, the PBL-line
obtained in parallel from the same patient (E.S.) contained only a low number of V-ß6 + T cell. This
would argue for V-ß6 + T cells in general do not have a proliferative advantage when cultured in IL-2,
although a higher number of paired TIL/PBL derived lines would need to be compared in order to make
a definite conclusion in this regard (work in progress).

The observation that T cells derived from ovarian carcinomas have a dominant expression of a certain V-
gene family could be explained by the assumption that T cells exposed to and possibly activated by tumor
associated or tissue specific antigens, would expand more easily than non-activated cells when cultured
with IL-2, thus leading to their predominance in vitro[17]. If the T cell activation in vivo would result from
stimulation with a limited set of antigenic peptides this should result in a relatively oligoclonal expansion.
When the oligoclonality of the V-ß6 + cells was analysed with a recently developed "J-ß ranking" method,
the V-ß6 + T cells in the two investigated TIL lines showed very (patient E.S.) or relatively (patient
M.M.) restricted clonality, which seemed to increase as a result of in vitro culture. Studies with a larger

number of patients is necessary to generalize these findings, and also to be able to study whether any particular combination of V-ß - J-ß would dominate.

In fresh tumor samples restricted TCR V-α expression has previously been observed in TILs of human melanomas and gliomas[18], as measured with PCR technique. With a similar experimental approach, we have recently failed to see any apparent restriction with regard to TCR V-α or -ß usage in tumor biopsies of ovarian carcinomas (Halapi et al, unpublished observation), agreeing with the conclusion that only upon in vitro culture of TIL is clonal dominance observed in this disease. Examination of genomic rearrangement of TCR DNA from patients with lung-cancer indicated the presence of oligoclonal T cells[19]. Other studies failed to detect unique rearrangement of the TCR in freshly isolated lymphocytes from cancer patients but oligoclonal populations could be detected after in vitro culture with IL-2[20,21]. In line with our studies, Karpati et al[22] have reported dominant expression of single V-ß products from IL-2 activated TIL in a murine model system. In this study, there was no relationship between dominance in TCR V-ß expression and tumor reactivity in vitro or in vivo. We also failed to detect specific proliferative or cytotoxic activity when we isolated V-ß6[+] lymphocytes from the TIL cultures described here (data not shown). Further studies e.g. of cytokine production are however necessary to analyse the possible functional role of the V-ß6[+] TIL cells. Regardless of this, however, the finding of restricted TCR repertoire in T cells expanded from ovarian carcinoma tissue but not from PBL of the same patient rewards further considerations when comparing therapeutical efficiency of IL-2 expanded TIL cells as compared to PBL derived LAK cells.

ACKNOWLEDGMENT : This study was supported by the Swedish Cancer Society and the Swedish Society for Medical Research. The authors also thank Drs. M. Brenner and H. DerSimonian for the mAb reactive with V-α12.1

REFERENCES

1. Rosenberg, S., Spiess, P., & Lafreniere, R. Science, 233, 1318-1321, 1986.
2. Kradin, R., Kurnick, J., Lazarus, D., Preffer, F., Dubinett, S., Pinto, C., Gifford, J., Davidson, E., Grove, B., Callahan, R. & Strauss, W. Lancet, 577-580, 1989.
3. Bukowski, R., Sharfman, W., Murthy, S., Rayman, P., Tubbs., Alexander, J., Budd, G., Sergi, J., Bauer, L., Gibson., et al. Cancer Res., 51, 4199-4205, 1991.
4. Dillman R., Oldham, R., Barth, N., Cohen, R., Minor, D., Birch,R., Yanelli, J., Maleckar, J., Sferruzza, A., et.al. Cancer, 68, 1-8, 1991.
5. Vanky, F. and Klein, E. Int.J.Cancer, 29, 547-553, 1982.
6. Whiteside, T., Heo, D., Johnson, J. & Herbeman R. CancerImmunol.Immunother., 26, 1-10, 1988
7. Lotzova, E., Savary, C., Freedman, R., Edwards, C. and Morris, M. Cancer Immunol Immunther., 31, 169-175, 1990
8. Nitta, T., Oksenberg, J., Narsing, A., & Steinman L. Science, 249, 672-674, 1990.
9. DerSimonian, H., Band, H. and Brenner, M.B. J.Exp.Med., 174, 639-648, 1991.
10. Chomczynski, P. & Sacchi, N. Anal. Biochem., 162, 156-159, 1987 .
11. Choi,Y., Kotzin, B., Herron, L., Callahan, J., Marrack, P. and Kappler, J. Proc.Natl.Acad.Sci., 86, 8941-8945, 1989.
12. Grunewald, J., Jeddi-Teharani, M., Pisa, E., Jansson, C-H., Andersson, R. and Wigzell, H. Subm.for publ.
13. Grunewald, J., Jansson, C.H. and Wigzell, H. Eur.J.Immunol., 21, 819-822, 1991.
14. Cozzarizza, A., Kahan, M., Ortolani, C., Franceschi, C. and Londei, M.
15. Beverly, P. Immunological memory i T cells. Curr.Opin.Immunol., 3, 355-360,1991
16. Bell, E. and Sparshott, S. Nature, 348, 163-166, 1990.
17. Mayer, T., Fuller, A., Fuller, T., Lazarovits A., Boyle L. and Kurnick, J. J.Immunol., 134, 258-264, 1985.
18. Nitta, T., Sato, K., Okumora, K. and Steinman, L. Int.J.Cancer, 49, 545-550, 1991.
19. Yoshino, I., Yano, T., Yoshikai, Y., Murata, M., Sugimachi, K., Kimura, G. and Nomota, K. Int.J.Cancer, 47, 654-658, 1991.
20. Belldegrun, A., Kasid, A., Uppenkamp, M., Topolian, S. and Rosenberg S. Science, 233, 1318-1321. 1986.
21. Fishleder, A., Finke, J., Tubbs, R. and Bukowski, R. J.Natn.Cancer.Inst., 82, 124-128, 1990.
22. Karpati, R., Banks, S., Malissen, B., Rosenberg, S., Sheard, M., Weber, J. and Hodes, R. J.Immunol., 146, 2043-2051, 1991.

26

COMPARISONS OF THE V-DELTA-1 EXPRESSING T-CELLS IN SYNOVIAL FLUID AND PERIPHERAL BLOOD OF PATIENTS WITH CHRONIC ARTHRITIS

Kalle Söderström[1], Anders Bucht[2], Thomas Hultman[3],
Mathias Uhlén[3], Ethel Nilsson[4], Alvar Grönberg[2],
Satish Jindal[5] and Rolf Kiessling[1]

1. Department of Immunology, Karolinska Institute, Stockholm
2. Department of Pharmacology, Kabi Pharmacia AB, Autoimmunity, Uppsala
3. Department of Biochemistry, The Royal Institute of Technology, Stockholm
4. Department of Rheumatology, Karolinska Hospital, Stockholm
5. Procept Inc., Cambridge, MA, USA

INTRODUCTION

In the hematolymphoid tissues, the majority of T-cells express an antigen receptor (TcR) composed of clonally variable αβ heterodimer. A minor but distinct fraction of peripheral blood (PB) T-cells express a rδ TcR heterodimer[1], which in healthy individuals make up to around 5% of the total T-cell population[2]. In several species, however not described in humans, rδ T-cell represent a major population in some epithelial tissues[3]. The receptor found on rδ T-cells in the periphery mainly utilize Vδ2[4]. A smaller fraction uses the Vδ1, which otherwise is predominantly expressed on rδ thymocytes[5]. Apart from the αβ T-cells, which have many Vα- and Vβ gene segments available for the construction of functional TcR, rδ T-cells have a only a few different Vr- and Vδ genes. Despite few V genes, the repertoire of rδ T-cells is as large or perhaps even larger than that of αβ cells[1]. This is mainly generated by an extensive junctional diversity, i.e the segments between V and J for the r-chain, and between V and D, and D and J for the δ-chain[6,7].

The specificity and the biological role of the rδ T-cells is largely unresolved. But recent discoveries show that they are able to recognize a heterogeneous array of ligands, such as bacterial antigens[8,9], which could indicate that they have a protective function in infectious diseases, such as leprosy[10]. In autoimmune diseases, such as RA, no clear answer has emerged on what possible role rδ T-cells could play, however rδ T-cell clones recognizing mycobacterial antigens have been obtained from the joint of a patient with RA[11]. Considering the fact that rδ cells have been shown to recognize various self antigens, including heat shock protein (hsp)[12], and that hsp are highly conserved through evolution, the potential role of rδ cells in autoimmune disease needs to be clarified.

When investigating the amount of fresh rδ[+] T-cells in synovial fluid (SF), elevated as well as normal - compared to peripheral blood - levels have been described[13-15]. When analyzing the different subsets of rδ T-cells, with subtype specific monoclonal antibodies a more striking difference has been shown with an elevated number of Vδ1[+] cells in the synovial fluid[15-17].

The enhanced response of synovial T-cells, compared to peripheral T-cells, to mycobacterial antigens such as BCG and particularly the 65 kD hsp, suggest that this, or related proteins such as endogenous hsp, may play an important role in the activation of synovial T-cells[15,18,19]. Although there is no clear evidence that rδ T cells is responding to hsp cell in SF, it is interesting to find that stimulation of synovial T-cells with mycobacterial antigen and maintained on IL2, or stimulated with IL2 only, lead to a significant expansion of rδ T-cells, with a predominance of Vδ1[+] cells[15].

T Lymphocytes: Structure, Function, Choices, Edited by F. Celada
and B. Pernis, Plenum Press, New York, 1992

In this study we have characterized the Vδ1 cell in the SF of patients with chronic inflammatory joint diseases. The data show that the Vδ1 cells with a preactivated memory type predominate among rδ T-cells in the joint. In contrast, the peripheral blood Vδ1 cells are of a naive unprimed type. These SF Vδ1 cells also tend to expand upon culture in IL2 alone. When studying the junctional diversity of the Vδ1 cell in IL2 expanded lines from SFMC and PBMC from four patients (two RA and two jRA), the result indicate that although there is an extensive heterogeneity in the junctional region of Vδ1 cells from most patients, it was possible to detect oligoclonally expanded Vδ1 cells in SF of one patient.

The existence of rδ T-cells with limited clonality could argue for their recognition of some highly conserved self-antigen, such as hsp. In line with this possibility, we have found that SFMC but not PBMC from some RA patients are able to respond specifically to human recombinant hsp 60.

MATERIAL AND METHODS

Patients: Twenty-four patients with chronic rheumatic diseases, mainly RA and two patients with reactive arthritis were included in the FACS analysis. For Vδ1 sequencing, IL2 expanded cell lines from 4 patients were used (AB and UL are females with RA; TG (male) and SB (female) have jRA).

Isolation of cells and culture conditions: The mononuclear cells were isolated by gradient centrifugation. After washing the cells were cultured in RPMI 1640 (GIBCO), with antibiotics, human serum, glutamine and recombinant IL2, as previously described[15]. The cells were expanded by feeding every 3-4 day, with this medium. Two to three weeks later the cells were harvested and stained.

Monoclonal antibodies and immunofluorescence: TCR-δ1 (anti-pan rδ TcR) and δTCS1 (anti-Vδ1-Jδ1/Jδ2) were purchased from T-cell sciences. Mab Leu-4 (anti-CD3), WT31 (anti-αβ TcR), Leu-18 (anti-CD45RA) and anti-HLA-DR were purchased from Becton Dickinson. Mab 4B4 (anti-CD29) was purchased from Coulter immunology and DAKO-UCHL1 (anti-CD45RO) from Dakopatts. Double direct staining was performed according to manufacturers recommendation. Double IF with δTCS1 was done by incubating 3×10^5 cells at 4°C for 30 min with optimal amount of mab, washed three times in cold PBS followed by addition of PE-conjugated goat anti mouse Ig F(ab)2'(Tago). Normal mouse serum was added 10 min before the cells were washed and incubated with the FITC-conjugated mab.

RNA preparation and cDNA synthesis: Total RNA from 2.5×10^6 cells using the NP-40 extraction method[22], was subjected to first strand cDNA synthesis using a TcR Cδ specific primer as described in ref. 21.

Polymerase chain reaction, subcloning and DNA sequencing: TcR Vδ1 cDNA was PCR amplified using a Vδ1 sense primer together with a Cδ antisense primer nested to the cDNA synthesis primer[23]. The Vδ1 primer included a Sph I site, and the C primer a Sal I site. PCR products were cleaved and purified using Geneclean kit (Bio 101) and ligated into a Sal I/Sph I cleaved puc18 plasmid. After E.coli transformation and blue/ white selection, positive clones were picked and further PCR amplified using puc18 primers, one of which was biotin- labelled. The DNA sequencing was done as described in ref. 24 and 25.

Proliferation assay and antigens: The time of culture and pulsing with tritiated thymidine was as in ref. 15. The human recombinant hsp 60 was kindly supplied by Dr. R. A. Young (Cambridge, MA, USA). The E.coli control have been made from E-coli transformed with the vector alone (without hsp gene) followed by the same strategy for protein induction and purification as used for hsp 60. The purity of hsp 60 is > 50%, based on SDS gel analysis. The same amount of E.coli proteins were present in the control as in the hsp 60 stimulation. The concentration of hsp 60 used was < 10 μg/ml.

RESULTS

The T-cell phenotypes of fresh and IL-2 cultured SFMC and PBMC

The phenotypic profile of fresh and short-term T-cell lines derived from stimulating SFMC and PBMC with IL-2 was analyzed in the FACS using anti-CD3, TCR-δ1 and δTCS-1. The majority of the cells in the fresh and IL-2 cultured SFMC and PBMC were CD3 positive T-cells (table 1).

The content of rδ+ cells was analyzed in SFMC and PBMC before and after IL-2 expansion in-vitro. Fresh SFMC had a slightly although not significantly higher amount of TCR-δ1+ cells compared to PBMC. A significantly higher proportion of TCR-δ1+ cells was seen in the IL-2 cultured compared to fresh SFMC (p<0.002, Mann-Whitney 2-tailed). Also the TCR-δ1+ cells in PBMC increase as a result of IL-2 culture, although this increase was not significant. Comparing the content of TCR-δ1+ between IL-2 expanded SFMC and PBMC, the amount was significantly higher in SFMC (0.02<p<0.05, Mann-Whitney 2-tailed), confirming our previously published results[15].

Table 1. T-cell phenotypes of fresh and IL-2 cultured SFMC and
PBMC from patients with RA.

Cell	Culture	% CD3[+]	% TCR-δ1[+a]	% δTCS-1[+b]
SFMC	-	76.7 ± 12.9(21)[c]	6.9 ± 6.4(19)	60.5 ± 24.1(6)
	IL2	62.8 ± 17.0(16)	33.1 ± 26.8(16)	73.2 ± 28.5(10)
PBMC	-	61.8 ± 13.1(19)	5.6 ± 5.8(18)	37.9 ± 19.0(5)
	IL2	82.1 ± 18.8(16)	12.1 ± 13.8(16)	36.9 ± 30.5(12)

[a]The % out of CD3[+], [b]the % out of TCR-δ1[+].
[c]Values represent the mean ± S.D., value in brackets
represent total number of patients studied.

We next analyzed the proportion of δTCS-1[+] among fresh and IL-2 cultured SFMC and PBMC. Only those cell samples which had more than 4% TCR-δ1[+] cells were analyzed. When comparing the percentage of TCR-δ1+ cells expressing δTCS-1+ among freshly isolated cells, a higher proportion of rδ T-cells expressed Vδ1 in SFMC compared to PBMC. This difference was not significant, which was probably due to the small number of observations. When analyzing IL-2 stimulated cell lines, SFMC had a significantly higher frequency of δTCS1[+] cells, as compared to PBMC derived ones ($0.002 < p < 0.02$, Mann-Whitney 2-tailed).

One possibility to explain why the Vδ1 cells derived from SF expanded more efficiently in IL-2 is that they are more activated in vivo than the PB derived Vδ1 cells. We have therefore analysed the expression of activation markers on paired samples of freshly isolated SFMC and PBMC from the same patient (table 2).

The expression of HLA-DR on δTCS1+ cells was significantly higher among SFMC than PBMC ($p < 0.01$, Wilcoxon 2-tailed). Also CD45RO expression was higher on the SFMC subset ($0.02 < p < 0.05$, Wilcoxon 2-tailed). Looking at the CD45RA expression, a lower expression was seen on the SFMC subset compared to PBMC ($p < 0.01$, Wilcoxon 2-tailed). No significant difference was seen for CD29, although the mean ± SD was higher in SFMC.

Taken together, this indicates that the Vδ1 subtype in SF is more activated, expressing high levels of HLA-DR and CD45RO, compared to its PB counterpart in the same individual. In PB, the resting type of Vδ1 cell predominates expressing CD45RA and low levels of HLA-DR.

Table 2. The phenotype of the Vδ1 T-cell in peripheral blood and synovial fluid.

Cell	CD29	HLA-DR	CD45RO	CD45RA
SFMC	467±135 (8)[a]	512±109 (9)	404±105 (9)	381± 91 (9)
PBMC	422± 68 (8)	358± 76 (9)	286±112 (9)	457±115 (9)

[a]Values indicate the mean fluorescence ± SD of the FITC labelled antibody, obtained when two gates (FL2 histogram gate; FSC vs SSC dotplot gate) were set on the FACscan in order to analyze only V-δ-1 PE-labelled cells. Number in parenthesis represent the total number of patients studied, with paired SF and PB data.

PCR amplification and Vδ1 DNA sequencing.

Vδ1 cDNA, was PCR amplified from IL2 expanded SFMC lines from 4 patients, and 3 PBMC lines derived from the same patients. In all cases PCR amplified products were clearly visible after ethidium bromide staining. After ligation to puc18 plasmid and E.coli transformation, several clones were picked from each sample and sequenced. In table 3, the frequency of translatable sequences in relation to the total number of sequenced clones is shown, as well as the frequency of clones with identical sequences among the translatable ones.

Patient	Cells	Translatable (%)	Identical (%)
UL	SFMC	18/21 (86%)	2/18 (11%)
UL	PBMC	10/18 (56%)	2/10 (20%)
SB	SFMC	10/11 (91%)	0/10 (0%)
SB	PBMC	16/19 (84%)	3/16 (19%),3/16 (19%)[a]
TG	SFMC	24/24 (100%)	23/24 (96%)
TG	PBMC	19/20 (95%)	0/19 (0%)
AB	SFMC	11/12 (92%)	2/11 (18%)

[a]Two different sequences found, each with 19% frequence.

The results show that although there is an extensive heterogeneity in the junctional region of Vδl cells from three patients, it was possible to detect oligoclonally expanded Vδl cells in SF but not in PB of one patient.

Proliferative response of SFMC and PBMC to human recombinant hsp 60 and E.coli control

The SFMC and PBMC from seventeen patients (16 RA, 1 jRA) were tested for the proliferative response to recombinant hsp 60 and its corresponding E.coli control (table 4). Five patients had SFMC which responded (stimulation index > 3), while PBMC from only one of the patients gave a weak response. None of the responders had jRA, which were shown to be the only hsp 60 responders in another study[21]. Some of the reactivity of SFMC against hsp 60 could however be accounted for by E.coli contaminants in this semipurified hsp 60 preparation, as the E.coli control preparation also yielded significant responses in 2 of the 5 SFMC cultures. The higher proliferative response of SFMC as compared to PBMC against the E.coli control confirms that SFMC are superior in responding to bacterial proteins other than hsp 60[26].

Table 4. Proliferative response of 5 RA patients to recombinant human hsp 60 and E.coli control.

Patient	Cell	Control	hsp60	E.coli
1.	SFMC	1722[a]	11.4[b]	5.6
	PBMC	775	1.4	1.2
2.	SFMC	1607	4.2	0.8
	PBMC	1367	1.3	0.4
3.	SFMC	734	3.2	2.7
	PBMC	736	0.8	0.5
4.	SFMC	1656	11.1	11.1
	PBMC	1180	0.9	1.3
5.	SFMC	1259	5.7	2.4
	PBMC	498	3.9	3.3

aThe cpm values obtained after pulsing the medium cultures with 3H-thymidine day 6.
bThe stimulation index derived from the formula: cpm with antigen/cpmmedium control

DISCUSSION

We have shown that IL2 expanded cell lines from SF had a marked increase of rδ cells as compared to unstimulated T-cells from fresh SFMC or fresh- and IL2 expanded PBMC. Interestingly, the Vδl subset represents a major fraction both in fresh- and IL-2 expanded SF rδ T-cells, while in PB constitutes a minority, confirming previously published data[15]. One possibility is that the Vδl T-cells selectively migrate to the joint. Similar mechanism may be used by intraepithelial lymphocytes in the human intestine, which have been shown to harbor a high proportion of Vδ1 cells[27]. Alternatively, some unknown stimulus causes activation and local expansion of Vδ1 cells either caused by a specific antigen or a recruitment due to inflammatory signals. It has been shown that rδ cells accumulate in granulomatous lesions in leprosy[10] and in blood and epithelial lining of the lung of sarcoidosis patients[28], as well as in the intestinal epithelium in patients with coeliac disease[29].

Although there are many data showing that rδ cells recognize bacterial antigens, and that SF T-cells from RA patients show high responses to bacterial antigens, there is little evidence to suggest that this response is mediated by rδ cells. Antigens such as hsp, which have a high degree of sequence homology between all species, may provide a link between infectious agents and autoreactivity[30]. SF T-cells from patients with chronic inflammatory arthritis often show a high proliferative response to bacterial hsp 65 and some (mainly jRA patients) also to the human homologue hsp 60[21]. In contrast, we have shown in this study that this response is not unique for jRA patients since all responders in this study had RA. The only patient with jRA in this study was a non-responder to the recombinant hsp 60.

The proliferative response to human hsp 60 has not been linked to any particular subset of T-cell in the joint of RA patients. Normal PB, when stimulated with mycobacterial antigens or with Burkitt lymphoma cells, which express hsp, leads to a selective expansion of Vδ2+ cells[31,32]. In contrast, lines derived from SFMC of RA patients, stimulated with these antigens and maintained on IL2 gave rise to lines consisting mainly of Vδ1+ cells. Thus it appears that SFMC may contain a unique population of Vδ1 cells which is preferentially expanded upon stimulation with bacterial antigens. The reason for this in vitro expansion may be that cells are primed in vivo. One indication of in vivo preactivated Vδ1 cells was the enrichment of these cells after culture in IL2 only. We did not analyse high affinity IL2 receptor expression on Vδ1 cells in this study. However IL2 receptor bearing rδ cells have previously been detected in rheumatoid SF but not in PB[33].

By FACS double staining we have demonstrated that the fresh SF Vδ1 cells belong to the activated "memory" type of cell, with a high expression of HLA-DR and CD45RO, in contrast to the PB Vδ1 cell, which is of a "naive" unprimed type expressing CD45RA and low amounts of HLA-DR. In normal individuals, PB Vδ1 cells have a low expression of CD45RO as compared to Vδ2 cells[34], which is interesting in regard to the ability of PB derived Vδ2 cells to recognize hsp. We did not analyze the state of activation of Vδ2+ cells in SF or PB, which needs to be done to further characterize these two sub-sets in regard to potential hsp recognition.

In the second part of this study we performed molecular analysis of the Vδ1 TcR of SF and PB cells from 2 RA and 2 jRA patients. PCR amplification of Vδ1 cDNA, cloning and automated solid phase sequencing made it possible to predict the amino acid sequence from a large number of cDNA clones. All patients had a high proportion of Vδ1 among their TcR rδ both in PBMC and SFMC, with exception of PBMC from patient AB (data not shown). Sequence analysis confirmed that the majority of Vδ1 cDNA clones utilize Jδ1 (96%) and rarely Jδ2 (1%) or Jδ3 (3%). Extensive diversity in junctional nucleotide- as well as amino acid composition in the junctional region was found in all samples but one, indicating a polyclonal population of Vδ1 expressing cells. This suggests that in general, IL2 expansion does not induce a selective outgrowth of certain T-cell clones. However, SFMC from one jRA patient revealed striking cDNA δ sequence homogeneity, which was reproduced by doing another cDNA synthesis from the same RNA preparation (data not shown). This patient had a low number of rδ cells in the SFMC and PBMC lines but the proportion of Vδ1 of total rδ cells was high in both lines (100%). Experiments are now in progress on fresh samples taken from this patient to determine if rheu-matoid SF may harbor Vδ1 cells originating from a small number of clones.

In conclusion, we have shown that Vδ1 cells of the rheumatoid SF have a high expression of activation markers, indicating a preactivated population in vivo. In addition, IL2 expansion of synovial cells yield a high proportion of rδ cells, in most samples expressing predominantly Vδ1. Furthermore, the Vδ1 T-cells seem to be polyclonal although in 1 out of 4 patients studied a high degree of clonality exist in the IL2 expanded SF cell line. Since the function of rδ cells as well as the recognized antigens and HLA restriction, is not well defined, it is difficult to make conclusions regarding the meaning of our find-ings. However it is reasonable to assume that the structure of TcR V-D and D-J junctions of the δ chain and V-J junctions of the r chain has an impact on antigen recognition and/or HLA binding. In that case, diversity in the junctional regions may indicate multiple peptide specificities, either by recognizing different antigens or by recognizing different epitopes on the same protein. Limited rδ cell specificity may be the case in one particular patient in this study.

245

Acknowledgement

This work was supported by the Swedish Medical Research Council, the Swedish National Association against Rheumatism and the Swedish Board for Technical Development.

REFERENCES

1. Davis, M. M., Bjorkman, P. J. Nature 1988. 334: 395.
2. Groh, V., Porcelli, S., Fabbi, M., Lanier, L. L., Picker, L. J., Anderson, T., Warnke, R. A., Bhan, A. K., Strominger, J. L., Brenner, M. B. 1989. J. Exp. Med. 1989. 169: 1277.
3. Allison, J. P., and Havran, W. L. 1991. Annu. Rev. Immunol. 9: 679.
4. Triebel, F., Hercend, T. Immunol. Today 1989. 10: 186.
5. Casorati, G., de Libero, G., Lanzavecchia, A., and Migone, H. N. J. Exp. Med. 1989. 170: 1521.
6. Chien, Y. H., Iwashima, M., Wettstein, D. A., Kaplan, K. B., Elliot, J. F., Born, W., Davis, M. M. Nature 1987. 330: 722.
7. Lafaille, J. J., DeCloux, A., Bonneville, M., Takagaki, Y., Tonegawa, S. Cell 1990. 59: 859.
8. Haregewoin, A., Soman, G., Hom, R. C., and Finberg, R. W. Nature 1989. 340: 309.
9. Abo, T., Sugawara, S., Seki, S., Fuji, M., Rikiishi, H., Takeda, K., and Kumagai, K. Int. Immunol. 1990. 2: 775.
10. Modlin, R.L., Pirmez, C., Hofman, F. M., Torigian, V., Uyemura, K., Rea, T. H., Bloom, B. R., and Brenner M. B. Nature 1989. 339: 544.
11. Holoshitz, J., Konig, F., Coligan, J. E., de Bruyn, J., and Strober, S. Nature 1988. 339: 226.
12. O'Brien, R. L., Happ, M. P., Dallas, A., Palmer, E., Kubo, R., and Born, W. K. Cell 1989. 57: 667.
13. Brennan, F. M., Londei, M., Jackson, A. M., Hercend, T., Brenner, M. B., Maini, R. N., and Feldmann, M. J. Autoimmunity 1988. 1: 319.
14. Réme, T., Portier, M., Frayssinoux, F., Combe, B., Miossec, P., Favier, F., Sany, J. Arthritis Rheum. 1990. 33:485.
15. Söderström, K., Halapi, E., Nilsson, E., Grönberg, A., van Embden, J., Klareskog, L., Kiessling, R. Scand. J. Immunol. 1990. 32: 503.
16. Sioud, M., Kjeldsen-Kragh, J., Quayle, A., Kalvenes, C., Waalen, K., Förre, Ö., and Natvig., J. B. Scand. J. Immunol. 1990. 31: 415.
17. Wood, N., Rittershaus, C., Kung, P. C., Poplonski, L., Purvis, J., Snow, K. M., Keystone, E. C. Arthritis Rheum. 1989. 32: S11.
18. Holoshitz, J., Klajman, A., Drucker, I., Lapidot, Z., Yaretsky, A., Frenkel, A., van Eden, W., Cohen, I. R. The Lancet 1986. ii: 305.
19. Res, P. C. M., Schaar, C. G., Breedveld, F. D. van Eden, W., Cohen, I. R., and de Vries, R. R. P. The Lancet 1988. ii: 478.
20. Jindal S., Harley C. B., Dudani, A. K., Gupta, R. S. Mol. Cell. Biol. 1989. 9: 2279.
21. De Graeff-Meeder, E. R., van der Zee, R., Rijkers, G. T., Schuurman, H. -J., Kuis, W., Bijlsma, J. W. J., Zegers, B. J. M., van Eden, W. The Lancet 1991. 337: 1368.
22. Gough, N. M. Anal. Biochem. 1988. 173: 93.
23. Bucht, A., Söderström, K., Hultman, T., Uhlén, M., Nilsson, E., Kiessling, R., and Grönberg, A. Eur. J. Immunol. in press.
24. Hultman, T., Bergh, S., Moks, T., and Uhlén, M. BioTechniques 1989. 10: 84.
25. Hultman, T., Ståhl, S., Hornes, E., and Uhlén, M. Nucl. Acids. Res. 1989. 17: 4937.
26. Res, P. C. M., Telgt, D., van Laar, J. M., Oudkerk Pool, M., Breedveld, F. C., de Vries, R. R. P. Lancet 1990. 336: 1406.
27. Halstensen, T. S., Scott, H., Brandzaeg, P. Scand. J. Immunol. 1989. 30: 665.
28. Balbi, B., Moller, D. R., Kirby, M., Holroyd, K. J., Crystal, R. G. J. Clin. Invest. 1990. 85: 1353.
29. Viney, J., Macdonald, T. T., Spencer, J. Gut 1990. 31: 841.
30. Lamb, J. R., Bal, V., Mendez-Samperio, P., Mehlert, A., So, A., Rothbard, J., Jindal, S., Young, R. A., Young, D. B. Int. Immunol. 1989. 1: 191.
31. Kabelitz, D., Bender, A., Prospero, T., Wesselborg, S., Janssen, O., Pechhold, K. J. Exp. Med. 1991. 173: 1331.
32. Fisch, P., Malkovsky, M., Kovats, S., Sturm, E., Braakman, E., Klein, B. S., Voss, S. D., Morriessey, L. W., De Mars, R., Welch, W. J., Bolhuis, R. L. H., Sondel, P. M. Science 1990. 250: 1269.
33. Kjeldsen-Kragh, J., Quayle, A., Kalvenes, C., Waalen, K., Förre, Ö., Natvig, J. B. Scand. J. Immunol. 1990. 32: 651.
34. Parker, C. M., Groh, V., Band, H., Porcelli, S. A., Morita, C., Fabbi, M., Glass, D., Strominger, J. L., Brenner, M. B. J. Exp. Med. 1990. 171: 1597.

27

BISPECIFIC MONOCLONAL ANTIBODY TARGETED CYTOTOXIC T LYMPHOCYTES CAN RECYCLE

J.A.C. Voorthuis, E. Braakman, C.P.M. Ronteltap,
N.E.B.A.M. van Esch, and R.L.H. Bolhuis

Dr. Daniel den Hoed Cancer Center, Rotterdam, The
Netherlands

INTRODUCTION

Bispecific monoclonal antibodies (bs-mAb) that recognize an activation site on cytotoxic T lymphocytes (CTL) on the one hand and an antigen on tumor cells on the other hand, have been used to bridge CTL to tumor cells. This bs-mAb mediated interaction activates the CTL that subsequently delivers a lethal hit to the target cell[1,2,3].

Antigen specific CTL have the capacity to recycle[4]. After delivery of a lethal hit to the conjugated target cell they dissociate from the target cell and may subsequently engage and lyse another target cell.

The question we addressed in this study was whether bs-mAb targeted CTL, like antigen specific CTL, have the capacity recycle.

MATERIALS AND METHODS

Bispecific Monoclonal Antibody

We have used the bs-mAb anti-CD3 X MOV18 (directed against a human ovarian carcinoma (OVCA) associated antigen (gp38)). The anti-CD3 X MOV18 bs-mAb producing trioma cell line was obtained by somatic fusion of MOV18 hybridoma cells with spleen cells from a BALB/c mouse immunized with a CD3 positive human T cell clone[5].

Recycling Assay

Cloned CTL were targeted with bs-mAb, washed to remove unbound bs-mAb and incubated for various periods of time (0 to 24 h) with or without OVCA cells. The CTL were then harvested and tested again for their ability to lyse ^{51}Cr labeled OVCA cells in a standard 4 h cytotoxicity assay, either in the presence or absence of soluble bs-mAb.

T Lymphocytes: Structure, Function, Choices, Edited by F. Celada
and B. Pernis, Plenum Press, New York, 1992

Flow Cytometric Analyses

Bs-mAb density on the CTL surface was quantified by staining aliquots with phycoerythrin (PE) conjugated goat-anti-mouse Ab. Indirect immunofluorescence analysis using anti-MOV18-idiotype Ab (kindly provided by Dr. M. I. Colnaghi, Istituto Nazionale Tumori, Milan, Italy) was employed to investigate whether the MOV18 antigen binding site of the bs-mAb was occupied by gp38. Immuno-fluorescence analysis was performed on a FACScan.

Measurement of Intracellular Calcium

CTL were loaded with the fluorescent calcium chelator indo-1 to measure the levels of cytoplasmic free ionized calcium ($[Ca^{2+}]_i$) after crosslinking of bs-mAb labeled CD3 complexes. Rabbit-anti-mouse Ab was used as crosslinking agent. Blue fluorescence intensity, corresponding with $[Ca^{2+}]_i$, was measured on a fluorimeter.

RESULTS AND DISCUSSION

In the absence of OVCA cells or in the presence of irrelevant MOV18⁻ tumor-target cells, bs-mAb targeted CTL retained their bs-mAb mediated lytic capacity for several days, as reported earlier[6]. However, 3 - 6 h incubation of bs-mAb targeted CTL with OVCA cells resulted already in a complete loss of targeted cytotoxic activity against OVCA cells. Nevertheless, bs-mAb was still present on the surface of CTL that were incubated in the presence of OVCA cells, although some dissociation occured with increasing duration of the pre-incubation period. Comparison of bs-mAb density on bs-mAb targeted CTL incubated either with or without OVCA cells revealed that the amount of bs-mAb on the surface of inactivated CTL should be sufficient to induce lysis of OVCA cells. Retargeting of the inactivated CTL by the addition of soluble bs-mAb restored the capacity of the CTL to lyse OVCA cells. Thus, the inactivated state can not be attributed to exhaustion of the cytotoxic machinery of the CTL.
To analyse whether the unability of bs-mAb targeted CTL to recycle was due to occupancy of the MOV18 antigen binding site, by the target cell antigen gp38, we employed anti-MOV18-idiotype Ab. The fluorescence intensity on inactivated CTL is only slightly decreased as compared with bs-mAb labeled CTL incubated without OVCA cells. This observation indicates that most MOV18 antigen binding sites are still accessible.
In order to examine whether the induction of the anergic state is at the level of the CD3/TCR signal transduction pathway, we performed calcium mobilization experiments. One of the early events following crosslinking of membrane receptors is the increase in $[Ca^{2+}]_i$. To examine the signal transduction capacity of inactivated bs-mAb targeted CTL we measured $[Ca^{2+}]_i$ levels. Crosslinking, by rabbit-anti-mouse Ab, of bs-mAb labeled CD3 complexes on CTL incubated without OVCA cells resulted in an increase in $[Ca^{2+}]_i$. However, no calcium mobilization was induced when bs-mAb labeled CD3 complexes on inactivated CTL were crosslinked. This indicates that inactivation of bs-mAb targeted CTL is due to inactivation of the signal transduction capacity of those bs-mAb labeled CD3 complexes that have been previously engaged in signal transduction.

248

CONCLUSIONS

The results clearly demonstrate that, like antigen specific CTL, bs-mAb targeted CTL are able to recycle when soluble bs-mAb is added. Our findings indicate that inactivation is neither due to occupancy of the MOV18 antigen binding site of the bs-mAb on the CTL surface, nor to dissociation of bs-mAb from the CTL surface, but rather suggest that it is due to inactivation of the signal transduction capacity of previously engaged CD3/TCR complexes.

ACKNOWLEDGEMENT

Supported by the Dutch Cancer Society "Koningin Wilhelmina Funds," Grant DDHK 88-5; and Sandoz Research Stichting (travelling-scholarship).

REFERENCES

1. U. D. Staerz, O. Kanagawa, and M. J. Bevan, Hybrid antibodies can target sites for attack by T cells, Nature, 314:628 (1985).
2. D. M. Segal, M. A. Garrido, P. Perez, J. A. Titus, D. A. Winkler, D. B. Ring, A. Kaubisch, and J. R. Wunderlich, Targeted cytotoxic cells as a novel form of cancer immunotherapy, Mol. Immunol. 25:1099 (1988).
3. J. Van Dijk, S. O. Warnaar, J. D. H. Van Eendenburg, M. Thienpont, E. Braakman, J. H. A. Boot, G. J. Fleuren, and R. L. H. Bolhuis, Induction of tumor-cell lysis by bispecific monoclonal antibodies recognizing renal-cell carcinoma and CD3 antigen, Int. J. Cancer 43:344 (1989).
4. M. V. Sitkovsky, Mechanistic, functional and immunopharmacological implications of biochemical studies of antigen receptor-triggered cytolytic T-lymphocyte activation, Immunol. Rev. 103:127 (1988).
5. A. Lanzavecchia and D. Scheidegger, The use of hybrid hybridomas to target human cytotoxic T lymphocytes, Eur. J. Immunol. 17:105 (1987).
6. E. Roosnek and A. Lanzavecchia, Triggering T cells by otherwise inert hybrid anti-CD3/antitumor antibodies requires encounter with the specific target cell, J. Exp. Med. 170:297 (1989).

Heat shock protein (HSP), 81, 241-245
Hemagglutinin, 126-128
    surface exposed hemagglutinin, 37
Human immunodeficiency virus (HIV),
        123, 195-199, 207-213
    envelope glycoproteins (gp41, gp120),
        195-199, 207-213
    isolate IIIB, 207
    pol gene of HIV, 198
    recognition of HIV antigen by human
        Th cells, 195-199
    V3 loop sequence of HIV envelope,
        207
Human ovarian carcinoma (OVCA), 235,
        247

IDDM, *see* Insulin-dependent diabetes
        mellitus
Idiopeptide, 103, 122-123
    immunogenic idiopeptide, 103
    cryptic idiopeptide, 103
Idiotope, 112, 122
Idiotype, 97, 105, 111, 121-131
    idiotype interactions, 105, 112, 121
Idiotypic determinant, *see* Idiotope
Idiotypic network, 111
Immunoglobulin (Ig), 111, 121
    crosslinking of B cell surface Ig, 61
    H and L chains, 98, 100, 123-131
    Ig superfamily, 86, 165
    IgG Fc, 113
    IgM, 101-103
    L-chains, free, 98
Ig complementarity determining regions
        (CDR), 1, 148
    CDR1-3, 1
    CDR2, 148
    CDR3, 18, 33, 148
I-J molecule, 135
Immune response genes, 98
Immunodominance, 196
Immunogenicity, 195-199, 207-213
Immunointervention, 188
Immunological tolerance, 61
Immunoregulation, 105, 111, 135
Immunosuppression, 153, 187-193
    immunosuppression by class II
        blockade, 187-193
    immunosuppressive agents, 188
    immunosuppressive procedures, 153

Insulin, 113, 123, 153
Insulin-dependent diabetes mellitus
        (IDDM), 113, 145, 153, 163,
        187
Intrathymic selection, 169
Interferon γ (IFNγ), 135, 170
Interleukins, 56, 63, 135, 170, 227, 235
    IL-1, 63
    IL-2, 56, 63, 115, 136, 170, 227, 235,
        242-244
    IL-3, 130, 170
    IL-4, 135
    IL-5, 136
    IL-6, 63
    IL-10, 135
Internal image, 103, 121

Junctional (J) regions 1, 17, 26, 244

Keyhole limpet hemocyanin (KLH), 124,
        137-138

L cells, 37
LAK cells, *see* lymphokine-activated
        killer cells
Latent help, 84
Leprosy, 245
Ligands, 30, 123, 146
    dual ligands, 30
    lectin-like ligands, 115
Luteinizing hormone, 82
Lymphokine-activated killer (LAK) cells,
        235
Lymphokines, 135
Lymphopenia, 153
Lysosomes, 108
Lysozyme system, 197

Major histocompatibility complex (MHC),
        1, 17, 37, 46, 71, 97, 108,
        126, 137, 154, 179, 187
    MHC blockade *in vivo*, 190
        inhibition of antibody responses by,
            191
    MHC blockers as therapeutic agents,
        187-193
    MHC binding motifs, 37, 40-43, 71
    MHC class I, 1, 17, 37, 46, 108
        synthesis of class I molecules, 108

Tumor-infiltrating lymphocytes (TIL), 235

"Universal" peptide, 4
Uveoretinitis, 113, 147

V domain, 97, 113, 121
V gene of TcR, *see* T cell antigen receptor
Vaccine, 75-85, 198, 207
  birth control vaccine, 82
  hepatitis B vaccine, 213
  HIV vaccine, 198-199, 207-213
  malaria sporozoite vaccine, 84
  recombinant DNA subunit vaccines, 207, 213
  second generation vaccines, 84
  vaccine design, 75

Variable region, 1, 113, 121, 146
  V region disease hypothesis, 146
  V region gene usage, 113-114
  V(D)J junctional regions, 1, 17, 26
  VDJ coding sequence, 126
V3 loop of HIV envelope, 207 (*see also* Human immunodeficiency virus)
Vesicular stomatitis virus (VSV), 38
Virus-like particles, 207

Yeast transposon Ty, 207